MECHANICAL &
ELECTRICAL SYSTEMS

Marc Schiler

This publication is designed to provide accurate and authoritative information in regard to the subject matter covered. It is sold with the understanding that the publisher is not engaged in rendering legal, accounting, or other professional service. If legal advice or other expert assistance is required, the services of a competent professional person should be sought.

President: Roy Lipner
Vice-President of Product Development and Publishing: Evan M. Butterfield
Editorial Project Manager: Michael J. Scafuri
Production Coordinator: Daniel Frey
Creative Director: Lucy Jenkins

Published by Kaplan AEC Architecture
a division of Dearborn Financial Publishing, Inc.®
A Kaplan Professional Company
30 South Wacker Drive
Chicago, IL 60606-7481
(312) 836-4400
http://www.ALSonline.com

Printed in the United States of America.

04 05 06   10 9 8 7 6 5 4 3 2 1

CONTENTS

Welcome to Kaplan Architecture! Kaplan recently acquired Architectural License Seminars (ALS), the oldest and most respected provider of Architect Registration Examination (ARE) study products. Over 50 percent of the registered architects in practice have used ALS material to prepare for their exam. Kaplan Inc. is the nation's leading provider of lifelong education, having served more than 30 million individuals over the past 60 years.

THE ARE

Kaplan AEC Architecture provides the only complete, centralized source for all nine divisions of the ARE. All 50 states, 5 territories, and participating Canadian provinces offer the uniform NCARB Architect Registration Examination (ARE). This exam consists of nine divisions:

- Pre-Design
- General Structures
- Lateral Forces
- Mechanical & Electrical Systems
- Building Design/Materials & Methods
- Construction Documents & Services
- Site Planning
- Building Planning
- Building Technology

The Site Planning, Building Planning, and Building Technology divisions require graphic solutions to vignette graphic problems, while the other six divisions consist of multiple-choice questions. The exams are all administered by computer. Candidates must pass all divisions of the ARE in order to become a registered architect. Those who do not pass a division of the exam may retake it after six months. For further details on the ARE, please visit the NCARB Web site at www.ncarb.org.

Kaplan AEC Architecture provides a variety of study material to help you prepare for the exam: self study courses, computer mock exams, paper mock exams, question & answer handbooks, video workshops, and a CD-ROM-based flash card system. All of our products can be ordered online at www.ALSonline.com, or by calling (800) 420-1429.

The ARE is not an easy exam! Although we cannot guarantee a passing grade, we **can** guarantee our material will better prepare you for the ARE.

Good luck on your examination, and in your professional career.

ACKNOWLEDGMENTS

Kaplan AEC is grateful for the cooperation of many experts, professionals, organizations, and associations in the creation of this book. The following images are reproduced with the permission of the copyright holders.

Copyright 1989, American Society for Heating, Refrigerating and Air-Conditioning Engineers, Inc. (www.ashrae.org). Reprinted by permission from 1989 ASHRAE Handbook of Fundamentals. This text may not be copied nor distributed in either paper or digital form without ASHRAE's permission.

Courtesy of Carrier Corporation

Used by permission of Pilkington North American, Inc.

Reprinted courtesy of BBN Technologies.

Portions of this document reproduce tables from the *2003 International Plumbing Code,* International Code Council, Falls Church, Virginia. Reproduced with permission. All rights reserved.

COURSE PURPOSE AND PROCEDURE

This home study course is a condensed and simplified review of Mechanical and Electrical Systems. Courses covering the other divisions of the ARE are also available, all of which have been prepared by the professional staff of ALS.

We advise that you set aside a definite amount of study time each week, just as if you were taking a lecture course, and carefully read all of the material. At the end of each lesson you will find a short review quiz. The quiz questions are intended to be straightforward and objective, and the answers and explanations are included to permit self-checking.

Following the last lesson, there is a Final Examination, which has been designed to simulate the style of the actual exam. Complete answers and explanations will be found on the pages following the examination, to permit self-grading.

OVERVIEW

Architects have usually been less than enamored with the subject of Mechanical and Electrical Systems, which control the environment in buildings. And yet, these systems are largely responsible for achieving the most basic objectives in architecture: the health, safety, comfort, and convenience of a building's occupants.

In addition, lighting and acoustics are important determinants of our perception of a building. All of the aesthetics of design are based on what we see; all the great architects understood that, and understood that it is light that primarily defines the forms and spaces of a building. The additional factor, a more subtle one, is acoustics. Even with our eyes closed, we can sense the vast difference between a Gothic cathedral and our living room. Understanding these

phenomena can only improve the quality of the buildings we design.

In recent years, environmental issues have assumed increased importance in design. We have learned that we cannot pollute the world around us indefinitely. In addition, the supply of fossil fuels is not inexhaustible, and many of the sources of those fuels lie outside of our control. Thus, energy conservation and our response to the environment remain major constraints in the design of buildings.

Further, when we consider how plumbing affects health, and the danger to life posed by fire, we can understand why it is necessary for the architect to comprehend plumbing and life safety systems, and why the exam tests that knowledge. While these are not prime determinants of form, they are necessary considerations.

This course covers all of these issues at a level appropriate to the exam candidate. For those who wish to study any of the topics further, we have provided a short bibliography. We suggest that you get started early, study at a steady pace, and use the glossary as necessary.

This edition conforms with the 1994 and 1997 editions of the *Uniform Building Code*.

This course was written by Marc Schiler. Mr. Schiler, who has a bachelor's degree from USC and a master's degree from Cornell, has written, lectured, and consulted extensively on environmental controls and energy in buildings. He is currently a Professor at USC.

Material on the topic of sustainable design was written by Jonathan Boyer, AIA. Mr. Boyer is a principal of the firm Boyer Associates Ltd in Chicago, Illinois. He is a graduae of the University of Pennsylvania (BA) and Yale University (MArch). His firm has focused on sustainable design and environmental planning for over thirty years, with projects throughout the United States.

You are now ready to start your course in *Mechanical and Electrical Systems*. We wish you success in your examination.

BASIC THERMAL PROCESSES

INTRODUCTION

If we look at the history of shelter, we find that the original concerns were primarily related to thermal comfort. In fact, the word "shelter" may be defined as protection from the elements. Architecture has addressed many other issues since then, but the original considerations still remain. In this lesson, we will discuss some of the basic concepts and formulas related to thermal comfort.

The thermal processes of a building are analogous to those of the human body, in which the skin keeps the heat in or out and the circulation systems regulate the temperature. The skin of a building and that of the body perform the same function. The circulation system of a building consists of the mechanical equipment. The form and exposure of the building's skin is critical, and is determined by the architect. The circulation system is determined by the mechanical engineer, and as long as the architect provides sufficient space and doesn't create any insoluble problems the building can be kept appropriately warm or cool. To understand form, skin and exposure, and to avoid creating insoluble problems, we will first address basic heat transfer processes.

BASIC PHYSICS OF HEAT TRANSFER

There is a difference between heat and temperature—they are related but are not the same. Temperature is a measure of stored heat energy, but temperature is never transferred, only heat energy is.

Furthermore, heat is sometimes transferred without a change in temperature (such as the melting of an ice cube). When transferred heat causes a change in temperature it is called *sensible heat*. When it causes a change of state (ice to water or water to vapor) it is called *latent heat*. Heat always flows from a hotter object to a cooler object.

Two objects at the same temperature may store different amounts of heat (concrete stores a lot more heat than carpet, for example). This storage capacity is called the *specific heat (C_p)* of a material, which is measured by comparing it to the storage capacity of water. A *British Thermal Unit (Btu)* is defined as the amount of heat energy required to raise 1 pound of water by 1 degree Fahrenheit. Thus specific heat is measured in terms of the number of Btu's required to change the temperature of a specific material by 1 degree Fahrenheit.

Radiation

Radiation is the method by which heat is transferred between two objects not in contact and not shielded from each other. When you stand in the sun, you are experiencing heat transfer from the sun by radiation, in spite of the fact that you are not touching it (conduction), nor are you above it (convection). Radiation is always taking place, but usually at a slow rate. All objects radiate at each other, even two people standing near each other. The *wavelength* of the radiation is based on the *temperature* of the object. Warm things radiate infrared, while really hot objects (i.e., red hot steel) begin to glow in the visible spectrum. If they get even hotter, they glow orange, and then white hot. The *rate* of radiative exchange is based on the surface *temperature* of the objects, the *viewed angle* and a property

TABLE 1.1—SPECIFIC HEAT (C_p)

Description	Density lb/ft^3	Conductivity[b] λ Btu·in./ h·ft^2·F	Conductance (C) Btu/h· ft^2·F	Resistance [c] (R) Per inch thickness (1/λ) h·ft^2· F/Btu	Resistance [c] (R) For thickness listed (1/C) h·ft^2· F/Btu	Specific Heat, Btu/lb· deg F
MASONRY UNITS						
Brick, common[i]	120	5.0	—	0.20	—	0.19
Brick, face[i]	130	9.0	—	0.11	—	
Clay tile, hollow:						
1 cell deep ... 3 in.	—	—	1.25	—	0.80	0.21
1 cell deep ... 4 in.	—	—	0.90	—	1.11	
2 cells deep ... 6 in.	—	—	0.66	—	1.52	
2 cells deep ... 8 in.	—	—	0.54	—	1.85	
2 cells deep ... 10 in.	—	—	0.45	—	2.22	
3 cells deep ... 12 in.	—	—	0.40	—	2.50	
Concrete blocks, three oval core:						
Sand and gravel aggregate ... 4 in.	—	—	1.40	—	0.71	0.22
... 8 in.	—	—	0.90	—	1.11	
... 12 in.	—	—	0.78	—	1.28	
Cinder aggregate ... 3 in.	—	—	1.16	—	0.86	0.21
... 4 in.	—	—	0.90	—	1.11	
... 8 in.	—	—	0.58	—	1.72	
... 12 in.	—	—	0.53	—	1.89	
Lightweight aggregate ... 3 in.	—	—	0.79	—	1.27	0.21
(expanded shale, clay, slate ... 4 in.	—	—	0.67	—	1.50	
or slag; pumice): ... 8 in.	—	—	0.50	—	2.00	
... 12 in.	—	—	0.44	—	2.27	

Note: Unless otherwise noted, all tables in this lesson are from the 1989 ASHRAE Handbook of Fundamentals and are reprinted by permission.

VISIBLE SPECTRUM
(SHORT WAVELENGTHS)

GLASS
HIGH VISIBLE τ
LOW INFRARED τ

INFRARED
(LONG WAVELENGTHS)

WARMED OBJECT
RERADIATES INFRARED

THE HEAT IS TRAPPED INSIDE THE
GREENHOUSE BY THE GLASS.

GREENHOUSE EFFECT

called *emissivity*. Surfaces with high emissivities radiate at a rate higher than those with low emissivities.

The *emissivity* (ε) of a surface is a property of the material, and is usually the same as the *absorptivity* (α) at any *given wavelength*. The simplest example in the visible spectrum is color. Black surfaces have higher emissivities and absorptivities than white or shiny surfaces. Black surfaces heat up rapidly and cool off rapidly. Shiny surfaces heat up more slowly, but stay hot longer. The emissivity and absorptivity are often different between the infrared and visible spectrum. *Selective surfaces* are surfaces which have a high absorptivity in one wavelength (usually solar) and low emissivity in another (usually infrared). This means that the material stores incoming *solar* radiation without releasing it as *infrared* (i.e., a good solar collector panel).

The foil sometimes used on the back of fiberglass insulation has a very low emissivity, which reduces the radiative transfer across the air space inside the wall.

You are probably more familiar with *transmissivity* (τ), which is the measure of how easily a material allows radiant energy to pass through it. We normally consider glass to be transparent, since it has a very high transmissivity in the visible spectrum. In the infrared region, however, it has a very low transmissivity, which is what causes the "*greenhouse effect.*" Sunlight (short wavelength and visible portion of spectrum) is transmitted through the glass into a building, causing the materials inside to heat up. When they heat up, they reradiate, but in the infrared spectrum. This reradiated energy does not pass through the glass and is therefore trapped inside the building. This is similar to the behavior of a selective surface, only now the selection is in terms of what passes through, rather than what is absorbed.

One of the current questions society must examine is the greenhouse effect in the upper atmosphere. As more carbon dioxide is released into the air, the rate at which the earth reradiates into space changes. Many scientists believe that this is increasing the average temperature of the earth. This could be quite dangerous, because such climatic changes would cause melting

HORIZONTALLY

UPWARD

DOWNWARD

CONVECTION

in the polar caps, thus raising the levels of the oceans. At the same time, the snow line would rise in most mountains, reducing the amount of water stored for following summers. This would result in worse flooding in winter and worse droughts in summer. The question bears further examination, beyond the scope of this course.

If you are trying to warm a building, the greenhouse effect is a great benefit, but if you are trying to keep the building cool, it's a good reason to avoid horizontal skylights!

The *viewed angle* of a surface depends on the size of the surface and your distance from it. When you stand very close to a meat freezer it occupies a large angle of view, and you lose a surprising amount of heat to it. When you stand across the room, however, it occupies a smaller angle of view and there is consequently much less heat exchange. The average radiant temperature of your surroundings is called the *Mean Radiant Temperature (MRT)*, and is independent of air temperature. For example, when you are skiing on a sunny day, the air temperature

TABLE 1.2—AIR SPACE RESISTANCES

Position of Air Space	Direction of Heat Flow	Air Space Mean Temp (F)	Temp Diff (deg F)	0.5-in. Air Space[c] Value of E[d,e]					0.75-in. Air Space[c] Value of E[d,e]				
				0.03	0.05	0.2	0.5	0.82	0.03	0.05	0.2	0.5	0.82
Horiz.	Up	90	10	2.13	2.03	1.51	0.99	0.73	2.34	2.22	1.61	1.04	0.75
		50	30	1.62	1.57	1.29	0.96	0.75	1.71	1.66	1.35	0.99	0.77
		50	10	2.13	2.05	1.60	1.11	0.84	2.30	2.21	1.70	1.16	0.87
		0	20	1.73	1.70	1.45	1.12	0.91	1.83	1.79	1.52	1.16	0.93
		0	10	2.10	2.04	1.70	1.27	1.00	2.23	2.16	1.78	1.31	1.02
		−50	20	1.69	1.66	1.49	1.23	1.04	1.77	1.74	1.55	1.27	1.07
		−50	10	2.04	2.00	1.75	1.40	1.16	2.16	2.11	1.84	1.46	1.20
45° Slope	Up	90	10	2.44	2.31	1.65	1.06	0.76	2.96	2.78	1.88	1.15	0.81
		50	30	2.06	1.98	1.56	1.10	0.83	1.99	1.92	1.52	1.08	0.82
		50	10	2.55	2.44	1.83	1.22	0.90	2.90	2.75	2.00	1.29	0.94
		0	20	2.20	2.14	1.76	1.30	1.02	2.13	2.07	1.72	1.28	1.00
		0	10	2.63	2.54	2.03	1.44	1.10	2.72	2.62	2.08	1.47	1.12
		−50	20	2.08	2.04	1.78	1.42	1.17	2.05	2.01	1.76	1.41	1.16
		−50	10	2.62	2.56	2.17	1.66	1.33	2.53	2.47	2.10	1.62	1.30
Vertical	Horiz.	90	10	2.47	2.34	1.67	1.06	0.77	3.50	3.24	2.08	1.22	0.84
		50	30	2.57	2.46	1.84	1.23	0.90	2.91	2.77	2.01	1.30	0.94
		50	10	2.66	2.54	1.88	1.24	0.91	3.70	3.46	2.35	1.43	1.01
		0	20	2.82	2.72	2.14	1.50	1.13	3.14	3.02	2.32	1.58	1.18
		0	10	2.93	2.82	2.20	1.53	1.15	3.77	3.59	2.64	1.73	1.26
		−50	20	2.90	2.82	2.35	1.76	1.39	2.90	2.83	2.36	1.77	1.39
		−50	10	3.20	3.10	2.54	1.87	1.46	3.72	3.60	2.87	2.04	1.56
45° Slope	Down	90	10	2.48	2.34	1.67	1.06	0.77	3.53	3.27	2.10	1.22	0.84
		50	30	2.64	2.52	1.87	1.24	0.91	3.43	3.23	2.24	1.39	0.99
		50	10	2.67	2.55	1.89	1.25	0.92	3.81	3.57	2.40	1.45	1.02
		0	20	2.91	2.80	2.19	1.52	1.15	3.75	3.57	2.63	1.72	1.26
		0	10	2.94	2.83	2.21	1.53	1.15	4.12	3.91	2.81	1.80	1.30
		−50	20	3.16	3.07	2.52	1.86	1.45	3.78	3.65	2.90	2.05	1.57
		−50	10	3.26	3.16	2.58	1.89	1.47	4.35	4.18	3.22	2.21	1.66
Horiz.	Down	90	10	2.48	2.34	1.67	1.06	0.77	3.55	3.29	2.10	1.22	0.85
		50	30	2.66	2.54	1.88	1.24	0.91	3.77	3.52	2.38	1.44	1.02
		50	10	2.67	2.55	1.89	1.25	0.92	3.84	3.59	2.41	1.45	1.02
		0	20	2.94	2.83	2.20	1.53	1.15	4.18	3.96	2.83	1.81	1.30
		0	10	2.96	2.85	2.22	1.53	1.16	4.25	4.02	2.87	1.82	1.31
		−50	20	3.25	3.15	2.58	1.89	1.47	4.60	4.41	3.36	2.28	1.69
		−50	10	3.28	3.18	2.60	1.90	1.47	4.71	4.51	3.42	2.30	1.71

Position of Surface	Direction of Heat Flow	Surface Emittance					
		Non-reflective ε = 0.90		Reflective ε = 0.20		Reflective ε = 0.05	
		h_i	R	h_i	R	h_i	R
STILL AIR							
Horizontal	Upward	1.63	0.61	0.91	1.10	0.76	1.32
Sloping—45 deg	Upward	1.60	0.62	0.88	1.14	0.73	1.37
Vertical	Horizontal	1.46	0.68	0.74	1.35	0.59	1.70
Sloping—45 deg	Downward	1.32	0.76	0.60	1.67	0.45	2.22
Horizontal	Downward	1.08	0.92	0.37	2.70	0.22	4.55
MOVING AIR (Any Position)		h_o	R	h_o	R	h_o	R
15-mph Wind (for winter)	Any	6.00	0.17				
7.5-mph Wind (for summer)	Any	4.00	0.25				

Surface	Reflectivity in Percent	Average Emittance ε	Effective Emittance E of Air Space	
			One surface emittance ε; the other 0.90	Both surfaces emittances ε
Aluminum foil, bright	92 to 97	0.05	0.05	0.03
Aluminum sheet	80 to 95	0.12	0.12	0.06
Aluminum coated paper, polished	75 to 84	0.20	0.20	0.11
Steel, galvanized, bright	70 to 80	0.25	0.24	0.15
Aluminum paint	30 to 70	0.50	0.47	0.35
Building materials: wood, paper, masonry, nonmetallic paints	5 to 15	0.90	0.82	0.82
Regular glass	5 to 15	0.84	0.77	0.72

may be cool, but the radiant energy coming from the sun, and the reflection of the sun off the snow, combined with a little exercise, will make you quite warm. There is a special device which may be used to measure MRT, which is called a *globe thermometer*.

Convection

Convection is the heat exchange process which occurs only in a fluid medium, such as air or a liquid. Hot air rising is an example of convection. Air expands when it is hot, which reduces its density and makes it lighter. The cool, heavier air falls, while the warm, lighter air rises.

Smoke rises in chimneys because it is warmer than the room air. The only material that expands when it gets colder is water, and only just before it freezes. That is fortunate indeed, otherwise ice would always form at the bottom of a lake rather than on the surface, and all the fish would die. (It would also be difficult to ice skate.)

Convection occurs in rooms all the time, especially in large atrium spaces, as well as inside wall cavities, such as between studs. Convection is the only means of heat transfer which is strictly directional. It never transfers heat *downward*. It can transfer heat *horizontally* by stirring the air, but not as rapidly as *upward*. When the top of a space is warmer than the bottom, and the hot air rises and stays there, it is called *stagnation*. The difference in pressure in a vertical space (positive or outward at the top and negative or inward at the bottom) is called the *stack effect*. The rising air tries to push out at the top, and it pulls air in behind it, down below. The stack effect can be significant in tall office towers, in which the elevator shafts can act like smokestacks.

Tables indicating the thermal resistances of air spaces have different values for the same thickness, depending on the orientation of the space (horizontal or vertical) and the direction of the heat flow (up or down). The orientation is more critical than the thickness.

The thin film of air which occurs next to a wall also provides a resistance, calculated as the inverse of the so-called *film coefficient (fi)*.

Conduction

Conduction is the heat transfer process which occurs when objects are in direct contact. Picking up a hot frying pan results in immediate (and painful) conduction. Conduction is not directional; there is no preference for up or down, only from hotter object to colder object. In buildings, conduction occurs inside a wall, transferring heat from inside to outside in cold climates primarily by the direct contact of the different layers in the wall. Each material has a different *conductivity (k)*, and *resistivity (r)*, which is the inverse of conductivity. Specific thicknesses of materials have calculated *conductances (C)* and *resistances (R)*. The resistance (R) is calculated from the thickness (x) and the conductivity (k), using the formula:

$$x/k = R$$

Insulation is often specified by the letter R followed by a number. For example, R - 19 insulation has a resistance of 19 (ft^2 °F hr / Btu). A complete wall assembly has a calculated "conductance," which represents all the interactions of the internal materials, including some radiation and convection, and which is called the *U value*. The U value is the reciprocal of the sum of the resistances (1 divided by the sum):

$$U = 1/(R_1 + R_2 + R_3 + \cdots + R_n)$$

Example #1

The plan section shown on page 8 is for a wall being considered for a heat loss condition (winter). What is the U value of the wall?

Solution:

We tabulate the resistances in column format. We will calculate the U value both at a stud and

TABLE 1.3—THERMAL RESISTANCES

Description	Density lb/ft³	Conductivity[b] (k) Btu•in./ h•ft²•F	Conductance (C) Btu/h• ft²•F	Resistance [c] (R) Per inch thickness ($1/k$) h•ft²• F/Btu	Resistance [c] (R) For thickness listed ($1/C$) h•ft²• F/Btu	Specific Heat Btu/lb • deg F
INSULATING MATERIALS						
Blanket and Batt[d]						
Mineral Fiber, fibrous form processed from rock, slag, or glass						
approx.[e] 3–4 in.	0.3–2.0	—	0.091	—	11[d]	
approx.[e] 3.5 in.	0.3–2.0	—	0.077	—	13[d]	
approx.[e] 5.5–6.5 in.	0.3–2.0	—	0.053	—	19[d]	
approx.[e] 6–7.5 in.	0.3–2.0	—	0.045	—	22[d]	
approx.[e] 9–10 in.	0.3–2.0	—	0.033	—	30[d]	
approx.[e] 12–13 in.	0.3–2.0	—	0.026	—	38[d]	
MASONRY MATERIALS						
Concretes						
Cement mortar	116	5.0	—	0.20	—	
Gypsum-fiber concrete 87.5% gypsum, 12.5% wood chips	51	1.66	—	0.60	—	0.21
Lightweight aggregates including expanded shale, clay or slate; expanded slags; cinders; pumice; vermiculite; also cellular concretes	120	5.2	—	0.19	—	
	100	3.6	—	0.28	—	
	80	2.5	—	0.40	—	
	60	..7	—	0.59	—	
	40	1.15	—	0.86	—	
	30	0.90	—	1.11	—	
	20	0.70	—	1.43	—	
Perlite, expanded	40	0.93	—	1.08	—	
	30	0.71	—	1.41	—	
	20	0.50	—	2.00	—	0.32
Sand and gravel or stone aggregate (oven dried)	140	9.0	—	0.11	—	0.22
Sand and gravel or stone aggregate (not dried)	140	12.0	—	0.08	—	
Stucco	116	5.0	—	0.20	—	
WOODS (12% Moisture Content)[o,p]						
Hardwoods						0.39
Oak	41.2-46.8	1.12-1.25	—	0.89-0.80	—	
Birch	42.6-45.4	1.16-1.22	—	0.87-0.82	—	
Maple	39.8-44.0	1.09-1.19	—	0.94-0.88	—	
Ash	38.4-41.9	1.06-1.14	—	0.94-0.88	—	
Softwoods						0.39
Southern Pine	35.6-41.2	1.00-1.12	—	1.00-0.89	—	
Douglas Fir-Larch	33.5-36.3	0.95-1.01	—	1.06-0.99	—	
Southern Cypress	31.4-32.1	0.90-0.92	—	1.11-1.09	—	
Hem-Fir, Spruce-Pine-Fir	24.5-31.4	0.74-0.90	—	1.35-1.11	—	
West Coast Woods, Cedars	21.7-31.4	0.68-0.90	—	1.48-1.11	—	
California Redwood	24.5-28.0	0.74-0.82	—	1.35-1.22	—	
PLASTERING MATERIALS						
Cement plaster, sand aggregate	116	5.0	—	0.20	—	0.20
Sand aggregate 0.375 in.	—	—	13.3	—	0.08	0.20
Sand aggregate 0.75 in.	—	—	6.66	—	0.15	0.20
Gypsum plaster:						
Lightweight aggregate 0.5 in.	45	—	3.12	—	0.32	
Lightweight aggregate 0.625 in.	45	—	2.67	—	0.39	
Lightweight agg. on metal lath 0.75 in.	—	—	2.13	—	0.47	
Perlite aggregate	45	1.5	—	0.67	—	0.32

between the studs. We will then calculate the weighted average, which will represent the U value of the wall as a whole. The air film

adjacent to the wall on both sides is always considered as a vertical layer because of its resistance to heat flow. The orientation of air

6" CONCRETE (140 PCF)

2 × 4 FURRING @ 16" O.C.
(1½" × 3½" ACTUAL DIMENSION)

R-11 FIBERGLASS BATT
(3" ACTUAL DIMENSION)

½" AIR SPACE

½" GYPSUM PLASTER BOARD

CALCULATION OF U VALUE

films and air space is vertical and the heat flow is horizontal. The winter case always assumes a 15 mph wind outside and a low average temperature in the wall. Since the air space is not the dominant resistance in the wall (the insulation is) the temperature drop across the air space will be small. The emissivity on each side of the air space is high because we assume there is only building paper on the fiberglass, and nothing special on the back of the gypsum board. If there were foil backing on both, the emissivity would be low.

Note: All of the following values come from Tables 1.2 and 1.3.

	R@ gap	R@ stud
Outside air film (assume wind = 15 mph)	.17	.17
Concrete (k = 9, x = 6" x/k = 6/9 = .67)	.67	.67
Stud (x/k = 3.5/1.0 = 3.5)	–	3.5
Fiberglass (glass batt)	11.00	–
Air space (50°F mean temp, 30°ΔT, ε = .82)	.90	–
1/2" gypsum board	.32	.32
Inside air film (still air, ε = .90)	.68	.68
$R_{tot} = \Sigma R =$	13.74	5.34
$U = 1/\Sigma R =$	.073	.187

Finding a weighted average for the U value is based on 1 1/2" of .187 (stud) and 14 1/2" of .073 (insulated gap) for every 16" of wall.

$$\frac{1.5(.187) + 14.5(.073)}{16} = 0.08366$$

$$= 0.08 \ Btuh/ft^2 °F$$

Although this problem is more complicated than any single problem you are likely to encounter on the exam, it serves to illustrate a number of factors we have discussed. Such a problem might appear on the exam broken up into several problems.

Latent Heat

When you get warm enough, you begin to perspire. The sweat evaporates, cooling the body. That is also a form of heat transfer caused by the change of state of the perspiration from the liquid state (water) to the vapor state (humidity). This is called a *phase change*, and either stores ("uses up") or releases energy.

In the case of evaporation, it "uses up" energy, which was excess to the body *(latent heat of evaporation)*. Ice melting in a glass of ice water "uses up" heat, keeping the drink cool *(latent heat of fusion)*. If we wanted to recreate the ice cube, all of that energy would have to be extracted from the water again (by refrigeration) so in that sense, the energy is *stored* in the water. It is now becoming typical in solar designs to store energy using phase change materials such as eutectic salts (which dissolve or crystallize in water) or special paraffins which melt or solidify at comparatively low temperatures.

This stores and releases solar energy (heat) without tremendous temperature variations, which is vastly preferable in buildings.

Removing the moisture from the air in hot humid buildings always requires a great deal of energy, because of the heat that must be extracted to get the moisture to condense out (change state). You may remember that it takes 1 Btu to change 1 pound of water 1 degree Fahrenheit. How many Btu's does it take to get one pound of water from freezing to boiling?

$$212°F - 32°F = 180°F$$

$$180°F \times 1 \text{ Btu/°Ib} = \textit{180 Btu}$$

Compare this with the amount it takes to boil (or evaporate) the same pound of water at 212°F into one pound of steam, still at 212°F. The latent heat of evaporation is 1,000 Btu's/pound for water.

$$1 \text{ lb. } H_2O_{liq} @ 212°F + \textit{1,000 Btu}$$

$$=> 1 \text{ lb. } H_2O_{vap} @ 212°F$$

It takes 1,000 Btu's just to evaporate a pound of water compared to 180 Btu's for the entire temperature change from freezing to boiling!

HEATING LOAD CALCULATIONS

The heating load of a building consists of the sum of all of the losses through the building skin. Given an understanding of the basic heat transfer processes, how do we apply these concepts to an actual building?

Conduction (q_c or HL_c)

The same basic formula is used for conduction whether it occurs through the roof, walls, windows or doors. The heat flow is the product of the conductance of the assembly (expressed by the U value), the temperature differential between inside and outside (ΔT), and the exposed surface area of wall, window or roof (A). The instantaneous version of this formula, which measures the flow at a given instant in time, is:

$$q_c = U(A)\Delta T$$

$$= U(A)(T_{inside} - T_{outside})$$

The conduction is expressed in Btu's per hour, which is a measure of energy flow rate per unit of time at any given instant. If that flow rate were to continue for an hour, the given number of Btu's would be transferred. This is often abbreviated as Btu/h or Btuh.

A similar formula is used for an energy flow rate over a long period of time:

$$q_c = U(A)24(DD)$$

where DD refers to a number of *degree days* and q_c refers to total Btu's. A degree day is a measure of how cold it has been at a given place over a given period of time. One degree day is defined as a day whose mean temperature is one degree below the reference temperature of 65°F. Two degree days may be a single day with a mean temperature of 63° (two degrees below 65°) or two days at 64°. The amount of energy necessary to heat the building is assumed to be the same in either case. All days with a mean temperature above 65°F are disregarded. The idea is that we can record the temperature and calculate the degree days for a particular locality over a period such as a month and calculate the total heat loss which would result. We can even keep records for an entire winter. A mild winter would be a 3,000 degree day winter, while a severe winter would be a 7,000 degree day winter.

The instantaneous version of q_c is used to determine the case at a particular moment, usually called the *design day*. This is a day colder than 98 percent of the days experienced in that climate. If the heating equipment and plant are sized to keep the building warm on that day, they would be sufficient for the other 98 percent as well.

The degree day version of the formula for q_c may be used to compare two alternatives over a longer period of time. Thus, it is possible to determine a *payback* period on an investment, for example, the number of years of the reduced energy costs it would take to pay for the increase in insulation.

Example #2

If the design day for Hepzibah, N.Y., is 10°F, and if we expect to maintain an interior temperature of 65°F, what will be the conducted heat loss through 200 ft² of the wall section of Example #1?

Solution:

The calculated U value of Example #1 was .08 Btuh/ft²°F.

$$q_c = U\,(A)\Delta T$$
$$= .08 \text{ Btuh/ft}^2\text{°F} \times (200\text{ft}^2)$$
$$\times (65\text{°F} - 10\text{°F})$$
$$= 880 \text{ Btuh}$$

Furthermore, if Hepzibah experienced a 6,600 degree day winter, how much heat would be lost through the wall that year?

$$q_c = U(A)24 \text{ DD}$$
$$= .08 \text{ Btuh/ft}^2\text{°F} \times (200\text{ft}^2)$$
$$\times 24 \text{ hr} \times 6,600 \text{ DD}$$
$$= 2,534,000 \text{ Btu or about}$$
$$2.5 \text{ million Btu's}$$

Conduction Below Grade

Although q_c theoretically applies to building elements below the ground surface, it becomes very difficult to determine what value to use for the outside temperature, since it varies with depth and moisture. It is therefore difficult to determine a reasonable value for ΔT. In addition, the loss through basement walls and floor slabs is comparatively low. The values for *below grade walls* are therefore taken from a table based on ground water temperature, which is usually assumed to be the same as the average annual air temperature. The *slab on grade* values are taken from a table which considers whether or not the slab edge is insulated. The total loss (q_s) is simply the area times the factor taken from the table.

Infiltration

All buildings leak air. There is a steady flow of air in and out, through cracks such as those between window sash and frame, and even through walls where there are sockets and switches or (heaven forbid) where there is sloppy construction.

CASEMENT   SLIDING   DOUBLE HUNG   HOPPER   AWNING

AUSTRAL   PIVOTED   FOLDING   JALOUSIE   FIXED

WINDOW TYPES VIEWED FROM OUTSIDE

The outside air which leaks in replaces internal air and must therefore be heated or cooled from the outside temperature to the desired inside temperature. The heat required is called the infiltration load (q_i). It is calculated in two steps:

1. Determine the amount of air infiltration.
2. Determine the amount of heating (or cooling) required to bring the air to the proper temperature.

The amount of air infiltration may by determined using the *air change method* or the *crack method*. The air change method requires that you know the number of air changes per hour in the building. For existing buildings this may be determined by measurement, but for new buildings this must be estimated. The air change method is also useful for very tight buildings, such as offices, where there might not be much infiltration, but which require some minimum number of air changes per hour anyway, for hygiene or code reasons. The amount of air (Q_{cfh}, expressed in cubic feet per *hour*) is determined by multiplying the building volume in cubic feet (V) by the number of air changes (N):

$$Q_{cfh} = N \times V$$

The crack method is based on the number of linear feet of crack or joint in all of the windows in the space under consideration, which might be one room or the entire building.

For example, a 3 foot × 6 foot window has 3 ft + 6 ft + 3 ft + 6 ft = 18 feet of crack. If it were a double hung window, which has a joint in the middle as well, that would add 3 feet for a total of 21 feet. The amount of infiltration per linear foot is determined from a table which considers wind speed and window type. This value may then be multiplied by the number of linear feet:

$$Q_{cfh} = LF \times CFH/lin.ft.$$

Finally, the amount of heating or cooling required may be calculated using the equation:

$$q_i = .018 \ (Q_{cfh})\Delta T$$

$$= .018 \ (Q_{cfh})(T_{inside} - T_{outside})$$

Total Heating Load

We now have all of the pieces and can combine them to calculate the total heating load:

$$q_{total} = q_c + q_s + q_i$$

Note that q_c may be composed of several different "sub" q_c's, one for each surface, i.e., q_c for the walls, q_c for the windows and q_c for the roof. You will often have to break up q_c for the walls into one q_c for each type of wall

section, since each wall section may have a different U value.

TEMPERATURE GRADIENTS

While the above formulas tell us what is happening to the inside of a building, they do not tell us what is happening to the inside of a wall. For example, why do pipes within walls freeze and burst when the room temperature is 65°F? The temperature of each point within a wall depends on the resistance of the layers up to that point compared with the total resistance of all the materials in the wall. For example, though the room temperature may be 65°F, the temperature inside part of the wall may be below 32°F. The equation for the temperature drop across a layer is:

$$\Delta T_{layer} = (R_{layer}/R_{total}) \times \Delta T_{total}$$

Therefore we can calculate the temperature at the boundary between two layers by calculating the sum of the temperature gradients to that point.

Example #3

Determine the temperatures inside the walls in Hepzibah, N.Y. on the same design day as in Example #2. Investigate only the section through the air space, since there are water pipes there, which should not be allowed to freeze.

6" CONCRETE (140 PCF)

2 × 4 FURRING @ 16" O.C.
(1½" × 3½" ACTUAL DIMENSION)

R-11 FIBERGLASS BATT
(3" ACTUAL DIMENSION)

½" AIR SPACE

½" GYPSUM PLASTER BOARD

Solution:

	R	ΔT	T	Boundary of Outside air
			10.00°F	
Outside air film	.17	(.17/13.74) × 55° = 0.68		Air film
			10.68°F------------	
Concrete	.67	(.67/13.74) × 55° = 2.68		Concrete
			13.36°F------------	
Fiberglass	11.00	(11.0/13.74) × 55° = 44.03		Fiberglass
			57.39°F------------	
Air space	.90	(.90/13.74) × 55° = 3.6		Air space
			60.99°F------------	
Gypsum board	.32	(.32/13.74) × 55° = 1.29		Gyp.brd.
			62.28°F------------	
Inside air film	.68	(.68/13.74) × 55° = 2.72		Air film
			65.00°F	
				Inside air

COOLING LOAD CALCULATIONS

There are a number of internal sources of heat in a building, which should be considered when sizing cooling equipment, and when calculating the conducted transfer through the walls. This is complicated since several heat transfer processes are involved simultaneously. (See CLTD.)

People (q_p)

The occupants of a building themselves comprise one of the sources of heat gain. This can range from a minor factor (two or three people in a house) to a dominant constraint (3,000 people in an auditorium). Not only is the number important, but so is the activity. A human being generates about 450 Btuh at rest, but as much as 2,500 Btuh when engaged in heavy work or athletic activity. The cooling load may be calculated by looking up the activity and

TABLE 1.6—BTUH FOR VARIOUS ACTIVITIES

Degree of Activity	Typical Application	Total Heat Adults, Male Btu/h	Total Heat Adjusted[b] Btu/h	Sensible Heat Btu/h	Latent Heat Btu/h
Seated at rest	Theater, movie	400	350	210	140
Seated, very light work writing	Offices, hotels, apts	480	420	230	190
Seated, eating	Restaurant[c]	520	580 [c]	255	325
Seated, light work, typing	Offices, hotels, apts	640	510	255	255
Standing, light work or walking slowly	Retail Store, bank	800	640	315	325
Light bench work	Factory	880	780	345	435
Walking, 3 mph, light machine work	Factory	1040	1040	345	695
Bowling[d]	Bowling alley	1200	960	345	615
Moderate dancing	Dance hall	1360	1280	405	875
Heavy work, heavy machine work, lifting	Factory	1600	1600	565	1035
Heavy work, athletics	Gymnasium	2000	1800	635	1165

multiplying by the appropriate number of people:

$$q_p = \text{No. of people} \times \text{Btuh/person}$$

Lighting (q_l)

Lighting equipment generates light, of course. But it also generates heat. In fact, incandescent lighting generates much more heat than light, and even light itself eventually gets absorbed and turns into heat. Thus, even efficient fixtures generate heat, directly in proportion to their wattage. This heat must be removed by the cooling equipment like any other load. The equation for the load is:

$$q_l = 3.4 \ W$$

where W = the wattage of the equipment.

Equipment (q_m)

Mechanical and electrical equipment produce heat as well, and the cooling load can be determined in one of several ways, depending

on the information available. Sometimes the Btuh is part of the equipment specification, such as with a stove or heater. Sometimes only the wattage is available (typewriter, hairdryer, or TV). In such cases, use the formula for q_l on the previous page. In some cases, the only information available is the horsepower of the equipment (i.e., a diesel generator). Most equipment denominated in horsepower does not run at full horsepower most of the time, and thus a discounting factor is applied. The commonly accepted equation is:

$$q_m = 1,500 \times \text{Bhp}$$

where Bhp is brake horsepower (the most common measurement). One Bhp is actually 2,545 Btuh.

Cooling Load Temperature Differential (CLTD or ETD)

Calculating heat gain through walls is complicated by the fact that peak gains usually occur in sunny weather, and often occur when there is a large diurnal (day to night to day) temperature swing. Under such circumstances, the thermal mass and storage capacity of the wall, as well as its color and orientation, are important factors (see the discussion of C_p and ε on pages 2 and 3). The formulas for q_c on page 9 do not take any of these factors into account, yet it is difficult to separate them into individual equations. As a result, they are often lumped together to get an "equivalent" temperature differential to use in the q_c equation. Thus the new equation is:

$$q_{CLTD} = U\ (A)\ CLTD$$

where CLTD is Cooling Load Temperature Differential, or:

$$q_{ETD} = U\ (A)\ ETD$$

where ETD is Equivalent Temperature Differential.

The complications arise in determining CLTD. There are three steps:

1. Determine the wall classification or group of the wall or roof section.

2. Determine the base CLTD from the time of day, wall type, wall orientation and wall or roof color.

3. Adjust the CLTD based on the temperature history of the last 24 hours.

All of the steps are straightforward except the last. The tables are based on an average *outside* temperature of 85°F (i.e., a temperature swing of 75 to 95°F). If the average temperature was 8 degrees higher (say the swing was 86 to 100°F for an average of 93°F), then you would have to add 8 degrees to the CLTD. Similarly, the assumed *interior* temperature is 78°F, and if you had set the thermostat at a cooler temperature,

you would have to increase the CLTD by the appropriate amount.

Example #4

Let's take our Hepzibah, N.Y. wall and find the peak heat gain on an August day in which the average outdoor temperature during the preceding 24 hours was 83°F. Assume that the wall faces south, and that it is painted a dark color. What is the peak heat gain and when does it occur?

Solution:

We don't know when we will experience the greatest CLTD, because we don't yet know how long the thermal mass will delay the peak load.

1. Find the wall group from Table 1.7. The 8" concrete with 3" of insulation is not shown in the table, but there is a listing for 8" concrete with 2" of insulation in group A. Since we have more insulation than the wall listed, we will use the U value of 0.08 calculated in Example #1, rather than the tabulated value. The mass of the listed wall, however, is very close to that of the wall in the problem.

2. From Table 1.8, we see that the peak CLTD is 20°F and occurs at 11:00 p.m.

3. The average outdoor temperature over the preceding 24 hours was given as 83°F, which is 2° below the reference temperature of 85°F for the CLTD table. We must therefore lower the CLTD by 2°, resulting in a CLTD of 18°F.

4. The peak gain occurs at 11:00 p.m. and is:

$$
\begin{aligned}
q_{CLTD} &= U\ (A)\ CLTD \\
&= .08\ Btuh/ft^2{}^\circ F \times 200\ ft^2 \times 18^\circ F \\
&= 288\ Btuh
\end{aligned}
$$

TABLE 1.7—CLTD WALL GROUPS

Group No.	Description of Construction	Weight (lb/ft²)	U-Value (Btu/h·ft²·F)	Code Numbers of Layers (see Table 8)
4-in. Face Brick + (*Brick*)				
C	Air Space + 4-in. Face Brick	83	0.358	A0, A2, B1, A2, E0
D	4-in. Common Brick	90	0.415	A0, A2, C4, E1, E0
C	1-in. Insulation or Air Space + 4-in. Common Brick	90	0.174-0.301	A0, A2, C4, B1/B2, E1, E0
B	2-in. Insulation + 4-in. Common Brick	88	0.111	A0, A2, B3, C4, E1, E0
B	8-in. Common Brick	130	0.302	A0, A2, C9, E1, E0
A	Insulation or Air Space + 8-in. Common brick	130	0.154-0.243	A0, A2, C9, B1/B2, E1, E0
4-in. Face Brick + (*H.W. Concrete*)				
C	Air Space + 2-in. Concrete	94	0.350	A0, A2, B1, C5, E1, E0
B	2-in. Insulation + 4-in. Concrete	97	0.116	A0, A2, B3, C5, E1, E0
A	Air Space or Insulation + 8-in. or more Concrete	143-190	0.110-0.112	A0, A2, B1, C10/11, E1, E0
4-in. Face Brick + (*L.W. or H.W. Concrete Block*)				
E	4-in. Block	62	0.319	A0, A2, C2, E1, E0
D	Air Space or Insulation + 4-in. Block	62	0.153-0.246	A0, A2, C2, B1/B2, E1, E0
D	8-in. Block	70	0.274	A0, A2, C7, A6, E0
C	Air Space or 1-in. Insulation + 6-in. or 8-in. Block	73-89	0.221-0.275	A0, A2, B1, C7/C8, E1, E0
B	2-in. Insulation + 8-in. Block	89	0.096-0.107	A0, A2, B3, C7/C8, E1, E0
4-in. Face Brick + (*Clay Tile*)				
D	4-in. Tile	71	0.381	A0, A2, C1, E1, E0
D	Air Space + 4-in. Tile	71	0.281	A0, A2, C1, B1, E1, E0
C	Insulation + 4-in. Tile	71	0.169	A0, A2, C1, B2, E1, E0
C	8-in. Tile	96	0.275	A0, A2, C6, E1, E0
B	Air Space or 1-in. Insulation + 8-in. Tile	96	0.142-0.221	A0, A2, C6, B1/B2, E1, E0
A	2-in. Insulation + 8-in. Tile	97	0.097	A0, A2, B3, C6, E1, E0
H.W. Concrete Wall + (*Finish*)				
E	4-in. Concrete	63	0.585	A0, A1, C5, E1, E0
D	4-in. Concrete + 1-in. or 2-in. Insulation	63	0.119-0.200	A0, A1, C5, B2/B3, E1, E0
C	2-in. Insulation + 4-in. Concrete	63	0.119	A0, A1, B6, C5, E1, E0
C	8-in. Concrete	109	0.490	A0, A1, C10, E1, E0
B	8-in. Concrete + 1-in. or 2-in. Insulation	110	0.115-0.187	A0, A1, C10, B5/B6, E1, E0
A	2-in. Insulation + 8-in. Concrete	110	0.115	A0, A1, B3, C10, E1, E0
B	12-in. Concrete	156	0.421	A0, A1, C11, E1, E0
A	12-in. Concrete + Insulation	156	0.113	A0, C11, B6, A6, E0
L.W. and H.W. Concrete Block + (*Finish*)				
F	4-in. Block + Air Space/Insulation	29	0.161-0.263	A0, A1, C2, B1/B2, E1, E0
E	2-in. Insulation + 4-in. Block	29-37	0.105-0.114	A0, A1, B3, C2/C3, E1, E0
E	8-in. Block	47-51	0.294-0.402	A0, A1, C7/C8, E1, E0
D	8-in. Block + Air Space/Insulation	41-57	0.149-0.173	A0, A1, C7/C8, B1/B2, E1, E0
Metal Curtain Wall				
G	With/without air Space + 1-in./ 2-in. 3-in. Insulation	5-6	0.091-0.2 30	A0, A3, B5/B6/B12, A3, E0
Frame Wall				
G	1-in. to 3-in. Insulation	16	0.081-0.1 78	A0, A1, B1, B2/B3/B4, E1, E0

RADIATION THROUGH WINDOWS (q$_r$ OR SHGF)

There are several methods for calculating the radiant gain through glass, also known as *insolation* (not to be confused with insulation).

CLTD does not really apply, since the glass transmits radiation and conducts heat separately, because it has neither time lag nor thermal mass.

TABLE 1.8—CLTD VALUES

North Latitude Wall Facing	0100	0200	0300	0400	0500	0600	0700	0800	0900	1000	1100	1200	1300	1400	1500	1600	1700	1800	1900	2000	2100	2200	2300	2400	Hr of Maximum CLTD	Minimum CLTD	Maximum CLTD	Difference CLTD
Group A Walls																												
N	14	14	14	13	13	13	12	12	11	11	10	10	10	10	10	10	11	11	12	12	13	13	14	14	2	10	14	4
NE	19	19	19	18	17	17	16	15	15	15	15	15	16	16	17	18	18	19	19	20	20	20	20	20	22	15	20	5
E	24	24	23	23	22	21	20	19	19	18	19	19	20	21	22	23	24	24	25	25	25	25	25	25	22	18	25	7
SE	24	23	23	22	21	20	20	19	18	18	18	18	19	20	21	22	23	23	24	24	24	24	24	24	22	18	24	6
S	20	20	19	19	18	18	17	16	16	15	14	14	14	14	15	16	17	18	19	19	20	20	20	20	23	14	20	6
SW	25	25	25	24	24	23	22	21	20	19	19	18	17	17	17	18	19	20	22	23	24	25	25	25	24	17	25	8
W	27	27	26	26	25	24	24	23	22	21	20	19	19	18	18	18	19	20	22	23	25	26	26	26	1	18	27	9
NW	21	21	21	20	20	19	19	18	17	16	16	15	15	14	14	14	15	15	16	17	18	19	20	21	1	14	21	7
Group B Walls																												
N	15	14	14	13	12	11	11	10	9	9	9	8	9	9	10	11	12	13	14	14	15	15	15	15	24	8	15	7
NE	19	18	17	16	15	14	13	12	12	13	14	15	16	17	18	19	19	20	20	21	21	20	20	20	21	12	21	9
E	23	22	21	20	18	17	16	15	15	15	17	19	21	22	24	25	26	26	27	27	26	26	25	24	20	15	27	12
SE	23	22	21	20	18	17	16	15	14	14	15	16	18	20	21	23	24	25	26	26	26	26	25	24	21	14	26	12
S	21	20	19	18	17	15	14	13	12	11	11	11	12	14	15	17	19	20	21	22	22	22	21	21	23	11	22	11
SW	27	26	25	24	22	21	19	18	16	15	14	14	13	13	14	15	17	20	22	25	27	28	28	28	24	13	28	15
W	29	28	27	26	24	23	21	19	18	17	16	15	14	14	14	15	17	19	22	25	27	29	29	30	24	14	30	16
NW	23	22	21	20	19	18	17	15	14	13	12	12	12	13	15	17	19	21	22	23	23	24	24	24	24	11	23	9
Group C Walls																												
N	15	14	13	12	11	10	9	8	8	7	7	8	8	9	10	12	13	14	15	16	17	17	17	16	22	7	17	10
NE	19	17	16	14	13	11	10	10	11	13	15	17	19	20	21	22	22	23	23	22	21	20	20	20	20	10	23	13
E	22	21	19	17	15	14	12	12	14	16	19	22	25	27	29	29	30	30	30	29	28	27	26	24	18	12	30	18
SE	22	21	19	17	15	14	12	12	12	13	16	19	22	24	26	28	29	29	29	29	28	27	26	24	19	12	29	17
S	21	19	18	16	15	13	12	10	9	9	10	11	14	17	20	22	24	25	26	25	25	24	22	20	20	9	26	17
SW	29	27	25	22	20	18	16	15	13	12	11	11	11	13	15	18	22	26	29	32	33	33	32	31	22	11	33	22
W	31	29	27	25	22	20	18	16	14	13	12	12	12	13	14	16	20	24	29	32	35	35	35	33	22	12	35	23
NW	25	23	21	20	18	16	14	13	11	10	10	9	10	11	12	13	15	18	22	25	27	27	27	26	22	10	27	17
Group D Walls																												
N	15	13	12	10	9	7	6	6	6	6	6	7	8	10	12	13	15	17	18	19	19	19	18	16	21	6	19	13
NE	17	15	13	11	10	8	7	8	10	14	17	20	22	23	23	24	24	25	25	24	23	22	20	18	19	7	25	18
E	19	17	15	13	11	9	8	9	12	17	22	26	29	31	32	32	32	31	30	28	26	24	22	18	16	8	33	25
SE	20	17	15	13	11	10	8	8	10	13	17	22	26	29	31	32	32	32	31	30	28	26	24	22	17	8	32	24
S	19	17	15	13	11	9	8	7	6	6	7	9	12	16	20	24	27	29	29	29	27	24	22	19	19	6	29	23
SW	28	25	22	19	16	14	12	10	9	8	8	8	10	12	16	21	27	32	36	38	38	37	34	31	21	8	38	30
W	31	27	24	21	18	15	13	11	10	9	9	9	10	11	14	18	24	30	36	40	41	40	38	34	21	9	41	32
NW	25	22	19	17	14	12	10	9	8	7	7	8	9	10	12	14	18	22	27	31	32	32	30	27	22	7	32	25
Group E Walls																												
N	12	10	8	7	5	4	3	4	5	6	7	9	11	13	15	17	19	20	21	23	20	18	16	14	20	3	22	19
NE	13	11	9	7	6	4	5	9	15	20	24	25	25	26	26	26	26	26	25	24	22	19	17	15	16	4	26	22
E	14	12	10	8	6	5	6	11	18	26	33	36	38	37	36	34	33	32	30	28	25	22	20	17	13	5	38	33
SE	15	12	10	8	7	5	5	8	12	19	25	31	35	37	37	36	34	33	31	29	26	23	20	17	15	5	37	32
S	15	12	10	8	7	5	4	3	4	5	9	13	19	24	29	32	34	33	31	29	26	23	20	17	17	3	34	31
SW	22	18	15	12	10	8	6	5	5	6	7	9	12	18	24	32	38	43	45	44	40	35	30	26	19	5	45	40
W	25	21	17	14	11	9	7	6	6	6	7	9	11	14	20	27	36	43	49	49	45	40	34	29	20	6	49	43
NW	20	17	14	11	9	7	6	5	5	5	6	8	10	13	16	20	26	32	37	38	36	32	28	24	20	5	38	33
Group F Walls																												
N	8	6	5	3	2	1	2	4	6	7	9	11	14	17	19	21	22	23	24	23	20	16	13	11	19	1	23	23
NE	9	7	5	3	2	1	5	14	23	28	30	29	28	27	27	27	27	26	24	22	19	16	13	11	11	1	30	29
E	10	7	6	4	3	2	6	17	28	38	44	45	43	39	36	34	32	30	27	24	21	17	15	12	12	2	45	43
SE	10	7	6	4	3	2	4	10	19	28	36	41	43	42	39	36	34	31	28	25	21	18	15	12	13	2	43	41
S	10	8	6	4	3	2	1	1	3	7	13	20	27	34	38	39	38	35	31	26	22	18	15	12	16	1	39	38
SW	15	11	9	6	5	3	2	2	4	5	8	11	17	26	35	44	50	53	52	45	37	28	23	18	18	2	53	48
W	17	13	10	7	5	4	3	3	4	6	8	11	14	20	28	39	49	57	60	54	43	34	27	21	19	3	60	57
NW	14	10	8	6	4	3	2	2	3	5	8	10	13	15	21	27	35	42	46	43	35	28	22	18	19	2	46	44
Group G Walls																												
N	3	2	1	0	-1	2	7	8	9	12	15	18	21	23	24	24	25	26	22	15	11	9	7	5	18	-1	26	27
NE	3	2	1	0	-1	9	27	36	39	35	30	26	26	27	27	26	25	22	18	14	11	9	7	5	9	-1	39	40
E	4	2	1	0	-1	11	31	47	54	55	50	40	33	31	30	29	27	24	19	15	12	10	8	6	10	-1	55	56
SE	4	2	1	0	-1	5	18	32	42	49	51	48	42	36	32	30	27	24	19	15	12	10	8	6	11	-1	51	52
S	4	2	1	0	-1	0	1	5	12	22	31	39	45	46	43	37	31	25	20	15	12	10	8	5	14	-1	46	47
SW	5	4	3	1	0	0	2	5	8	12	16	26	38	50	59	63	61	52	37	24	17	13	10	8	16	0	63	63
W	6	5	3	2	1	1	2	5	8	11	15	19	27	41	56	67	72	67	48	29	20	15	11	8	17	1	72	71
NW	5	3	2	1	0	0	2	5	8	11	15	18	21	27	37	47	55	55	41	25	17	13	10	7	18	0	55	55

The formula for q_c can be used for the conducted gain. Radiation can be calculated by finding the intensity of the direct sun on the surface of the glass, and then multiplying by the glass area and the percentage transmitted. The percentage transmitted compared to that which is transmitted by clear glass is called the *shading coefficient (SC)* and is somewhat similar to the transmissivity. However, half of the radiation absorbed is assumed to reradiate into the space,

TABLE 1.9—SHADING COEFFICIENTS

A. Single Glass

Type of Glass	Nominal Thickness[b]	Solar Trans.[b]	Shading Coefficient	
			$h_0 = 4.0$	$h_0 = 3.0$
Clear	1/8 in.	0.86	1.00	1.00
	1/4 in.	0.78	0.94	0.95
	3/8 in.	0.72	0.90	0.92
	1/2 in.	0.67	0.87	0.88
Heat Absorbing	1/8 in.	0.64	0.83	0.85
	1/4 in.	0.46	0.69	0.73
	3/8 in.	0.33	0.60	0.64
	1/2 in.	0.24	0.53	0.58

B. Insulating Glass

Clear Out, Clear In	1/8 in.[c]	0.71[e]	0.88	0.88
Clear Out, Clear In	1/4 in.	0.61	0.81	0.82
Heat Absorbing[d] Out, Clear In	1/4 in.	0.36	0.55	0.58

[a] Refers to factory-fabricated units with 3/16, 1/4 or 1/2-in. air space or to prime windows plus storm sash.
[b] Refer to manufacturer's literature for values.
[c] Thickness of each pane of glass, not thickness of assembled unit.
[d] Refers to gray, bronze and green tinted heat-absorbing float glass.
[e] Combined transmittance for assembled unit.

as well. Therefore the tabulated shading coefficient values are different from those for transmissivity (see Table 1.9). Generally speaking, clear glass lets in a lot of light and heat, tinted glass lets in less light and some heat, and reflective glass lets in less light and less heat. Unfortunately, it is not always desirable to let in less light, because more artificial lighting may be required, which in turn generates more heat. It is best to stop the *direct* sunlight *outside* the building, using fins, overhangs, or awnings. The diffuse daylight which remains provides natural lighting and contributes much less heat. If direct sunlight is blocked inside the glass, with a Venetian blind or a drape, all of the heat that gets absorbed has already been admitted to the space. If it is necessary to use internal shielding, the more reflective or light colored the better.

The formula for radiant gain (sometimes called *solar heat gain factor* or just *solar factor*) is:

$$q_r \; (= SHGF = SF) = S_g \, (SC) \, A$$

where

S_g = the intensity (in Btuh/ft^2) on a surface area in a given orientation

SC = the shading coefficient

A = the area exposed to direct sunlight.

We will discuss solar gain further in the next lesson.

TABLE 1.10—SOLAR INTENSITY (S_G) FOR 40° NORTH LATITUDE

Date	Solar Time	Direct Normal Btuh•ft²	N	NNE	NE	ENE	E	ESE	SE	SSE	S	SSW	SW	WSW	W	WNW	NW	NNW	HOR	Solar Time
Mar 21	0700	171	9	29	93	140	163	161	135	86	22	8	8	8	8	6	8	8	26	1700
	0800	250	16	18	91	169	218	232	211	157	74	17	16	16	16	16	16	16	85	1600
	0900	282	21	22	47	136	203	238	236	198	128	40	22	21	21	21	21	21	143	1500
	1000	297	25	25	27	72	153	207	229	216	171	95	29	25	25	25	25	25	186	1400
	1100	305	28	28	28	30	78	151	198	213	197	150	77	30	28	28	28	28	213	1300
	1200	307	29	29	29	29	31	75	145	191	206	191	145	75	31	29	29	29	223	1200
	HALF DAY TOTALS		114	139	302	563	832	1035	1087	968	694	403	220	132	114	113	113	113	764	
Jun 21	0500	22	10	17	21	22	20	14	6	2	1	1	1	1	1	1	2	3	3	1900
	0600	155	48	104	143	159	151	121	70	17	13	13	13	13	13	13	14	40	1800
	0700	216	37	113	172	205	207	178	122	46	22	21	21	21	21	21	21	97	1700
	0800	246	30	85	156	201	216	199	152	80	29	27	27	27	27	27	27	153	1600
	0900	263	33	51	114	166	192	190	161	105	45	33	32	32	32	32	32	201	1500
	1000	272	35	38	63	109	145	158	148	116	69	39	36	35	35	35	35	238	1400
	1100	277	38	39	40	52	81	105	116	110	88	60	41	39	38	38	38	260	1300
	1200	279	38	38	38	40	41	52	72	89	95	89	72	52	41	40	38	38	267	1200
	HALF DAY TOTALS		253	470	734	941	1038	999	818	523	315	236	204	191	188	187	186	188	1126	
Dec 21	0800	89	3	3	8	41	67	82	84	73	50	17	3	3	3	3	3	3	6	1600
	0900	217	10	10	11	60	135	185	205	194	151	83	13	10	10	10	10	10	39	1500
	1000	261	14	14	14	25	113	188	232	239	210	146	55	15	14	14	14	14	77	1400
	1100	280	17	17	17	17	56	151	217	249	242	198	120	28	17	17	17	17	104	1300
	1200	285	18	18	18	18	19	89	178	233	253	233	178	89	19	18	18	18	113	1200
	HALF DAY TOTALS		52	52	56	146	374	649	822	867	775	557	276	94	53	52	52	52	282	

Solar Heat Gain Factors, Btuh•ft²

SUMMARY

In this lesson we have discussed the different modes of heat transfer through the building skin. Heating loads are calculated using:

$$q_{tot} = q_c + q_s + q_i$$

Cooling loads are calculated using:

$$q_{tot} = q_p + q_m + q_i + q_{CLTD} + q_r$$

There are several other concepts we have touched upon along the way. Mean radiant temperature is calculated based on the view angle and comparative temperatures. The temperature gradient inside of the wall is calculated based on comparative resistances and the overall temperature drop.

Finally, we have introduced a number of new terms, which may be the most important of all. Do you understand emissivity (ε), sensible heat, latent heat, specific heat (C_p), phase change, resistance and U value?

Take the quiz at the end of this lesson and when you are fairly confident about the information, proceed to the next lesson. Please make sure you go back to review anything you missed. Lessons One through Three are sequential, and you should therefore understand one before proceeding to the next.

LESSON 1   QUIZ

1. Insolation is an example of
   A. radiation.
   B. conduction.
   C. convection.
   D. latent heat transfer.

2. Which of the following is dependent on orientation?
   A. Radiation
   B. Conduction
   C. Convection
   D. Latent heat transfer

3. The mean radiant temperature relates to
   A. air temperature.
   B. surrounding surface temperatures.
   C. body temperature.
   D. all of the above.

4. CLTD approximates the effects of which of the following?
   I. Radiation
   II. Conduction
   III. Convection
   IV. Latent heat

   A. I and II   C. III and IV
   B. II and III   D. I and III

5. The resistance of a layer of construction in a wall relates to which of the following?
   I. The temperature gradient within the wall
   II. The conduction through the wall
   III. The conductivity of the material composing the layer

   A. I only   C. III only
   B. II only   D. I, II, and III

6. A degree day is directly related to
   I. the weather in an area over a period of time.
   II. a reference temperature.
   III. a design day.

   A. III only   C. I and II
   B. II and III   D. I and III

7. Infiltration is calculated using which method(s)?
   I. Design day method
   II. Air change method
   III. Crack method

   A. I only   C. III only
   B. II only   D. II and III

8. A calculation of heat loss includes
   A. q_c, q_{CLTD}, q_p and q_m.
   B. q_c, q_{CLTD}, q_s and q_l.
   C. q_c, q_s and q_i.
   D. q_s, q_i and q_r.

9. Using the *design day* temperatures in the q_c formula gives us information for which of the following?

    A. Equipment sizing

    B. Life cycle costing

    C. Latent heat values

    D. Yearly weather versus a reference temperature

10. Which of the following is contributed to the environment by a tea pot boiling on a stove?

    A. Sensible heat

    B. Convection

    C. Latent heat

    D. All of the above

2

COMFORT, CLIMATE, AND SOLAR DESIGN

INTRODUCTION

We learned about the basic thermal processes in the first lesson. We will now take that information and add to it certain ideas about what conditions are comfortable thermally, and some tools used to study and express different climates and human comfort ranges. After studying various climates, we will look at the shape traditional buildings take in response to those climates, and why. We will learn the basics of solar position and orientation. Finally, we will distill some old ideas out of the vernacular design, and add some new ideas and developments, which will result in what is generally called passive solar design.

HUMAN COMFORT RANGES AND ZONES

An Englishman is not comfortable in the same range of temperatures as a Tahitian. Human comfort ranges vary depending on culture, recent exposure and even health and age. However, the same factors always come into play. Let's begin by studying what those factors are.

Metabolism

In our discussion of cooling loads we considered the fact that humans generate somewhere between 450 and 2,500 Btuh. That heat must be dissipated from the human body by the same heat transfer mechanisms that remove heat from buildings. If that heat is not removed, body temperature begins to rise, which is uncomfortable at first, then unhealthy, and finally, even fatal.

In colder temperature ranges, radiation, conduction, evaporation and convection remove heat at a steady and rapid rate. We are warmer than our surroundings and the heat loss is much greater than the gain. In order to limit the heat loss, the body tends to close down the pores, which reduces evaporation and latent heat transfer. It raises goose bumps and contracts the capillaries, reducing blood flow near the skin, which reduces skin surface temperature (thus reducing radiative loss and conductive loss). If necessary, we put on additional layers of clothing, providing insulation and trapping convection.

As the temperature rises, the heat loss due to conduction, convection, and radiation slows down, and the body begins to adjust. First the pores open up, the blood flow near the surface increases, and finally, we begin to sweat. The moisture evaporates, using up the latent heat of evaporation and removing it from the body.

When the temperature of the environment reaches 98.6°F, all heat loss from conduction, convection, and radiation ceases. When the temperature rises any further, the direction of the heat flow reverses, since the body tries to remain at 98.6°F. At this point, the only mechanism still removing heat from the body is the evaporation of sweat and of moisture from the lungs and sinuses. This is why humidity is such a key factor. If the air around us is already at 90 percent relative humidity, the sweat will hardly evaporate. To understand the relationship of all of these factors, we will use a single chart called the *psychrometric chart*. This chart is immensely useful and works for just about everything engineers like to calculate, far beyond simple questions of human comfort.

The Psychrometric Chart

The psychrometric chart is a graph which shows the air at different temperatures and different humidities. You can also graph the total amount of energy stored in the air (sensible heat + latent heat) on the same chart, and the term typically used for that combined storage is called *enthalpy*. To cool and simultaneously dehumidify air requires removing both forms of stored heat, so we can just look at the change in enthalpy.

Lines representing the dry bulb temperature (and constant stored sensible heat) run vertically on the chart. Lines of wet bulb temperature (and constant stored enthalpy) run diagonally from the lower right to the upper left. The wet bulb temperature is the temperature measured using a thermometer with a wet sock on the bulb so that the rate of evaporation is taken into account. Dry air results in a large *wet bulb depression,* the difference between the dry bulb and the wet bulb temperatures. The thermometer with a sock is called a *psychrometer.* If it is

PSYCHROMETRIC CHART

Reprinted by permission of the Carrier Corporation.

SWING IN CIRCULAR MOTION

WET SOCK

SLING PSYCHROMETER

swung around in the air manually to get air movement, it is called a *sling psychrometer*.

The amount of water contained by the air is represented by the horizontal lines, either in *grains of moisture per pound of air*, or in *pounds of moisture per pound of air*, or in *pounds of moisture per 1,000 cubic feet of air* (even engineers don't agree on everything). *Relative humidity* is shown by curved lines which run from lower left to upper right. Note that a constant amount of water in the air does *not* represent a constant relative humidity (RH). Relative humidity is defined as the *percentage* of complete *saturation* (how much water is in the air at a given temperature compared to how much the air could hold at that temperature). Air can hold much more water when it is warm than when it is cold. So 0.010 lbs. of H_2O per lb. of air represents 90% RH at 60°F but only 20% RH at 105°F. This is why there is condensation on the outside of a cold glass of water or iced tea. As the warm, humid air gets cooled by the glass, it can't hold as much moisture, so some of it condenses out. This is also why we *always* put a vapor barrier on the *warm* side of the insulation in a wall. The temperature drop across the insulation will result in the air being able to carry less moisture. If we allowed moist air from the warm side (i.e., from inside the house) to

penetrate into the insulation, the moisture would condense out *within* the insulation, reducing its resistance, and in many cases causing building materials to deteriorate.

We can determine roughly what combinations of air temperature and relative humidity are comfortable for most people. We can outline that range of combinations on the psychrometric chart. That range of combinations is called the *comfort range* or *comfort zone*. In the United States, for people doing calm or sedentary work, and who are lightly clothed, it varies from 65°F to about 78°F and from 25% RH to about 75% RH. The higher RH values require slightly lower temperatures. For example, 75% RH is only pleasant up to 73°F. But at 25% RH, 78°F is just fine.

These comfort ranges are affected by factors which do not show on the psychrometric chart, for example, mean radiant temperature. We therefore use the term *effective temperature* for a combination of the ambient air temperature (dry bulb) and the MRT. If the MRT is high, then the comfort zone shifts to lower ambient air temperatures to make up for the difference, and when the MRT is low the comfort zone shifts to higher air temperatures.

Similarly, the local velocity of the air has an effect. When the air is moving, it carries away heat and moisture more rapidly. Thus, moving air shifts the comfort zone to slightly higher temperatures. If you sit in front of a fan, you can tolerate the heat better, even though you have not changed the temperature or the relative humidity.

SOLAR DESIGN

There are two environmental factors to which the design must respond: the sun and the climate. We will discuss the sun in sufficient detail for you to understand the basic concepts,

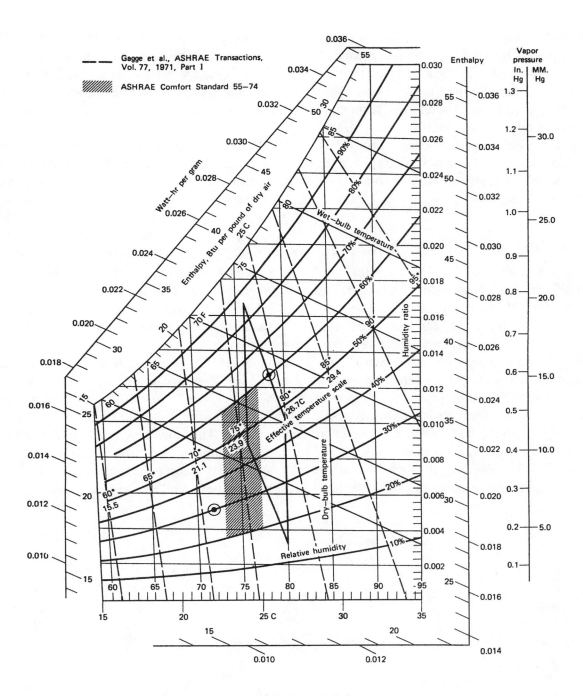

COMFORT ZONE

Reprinted from *Handbook of Fundamentals* by permission of ASHRAE.

and to handle any questions that the exam is likely to pose. This subject is worth a great deal more attention, and you are encouraged to study it further, because it is part of the global issue of energy. How you, as an architect, respond to the sun and other design factors will affect energy usage for decades to come. The following section should only be a beginning. For further study, check the references in the bibliography.

Solar Angles

The earth orbits the sun, and also rotates about its own axis, giving us day and night. The axis about which the earth rotates is not perpendicular to the plane of the orbit, but tilted at 23.5°. This means that part of the year the North Pole faces away from the sun, and part of the year it leans toward the sun. The seasons are caused by this changing *tilt* of the earth with respect to the sun, not by the changing *distance* of the earth from the sun. When it is winter in the Northern Hemisphere, it is summer in the Southern Hemisphere. The tilt of the North Pole in relation to the position of the sun is called

the *declination* angle (δ). On December 21 (*winter solstice*), the declination is –23.5°, and on June 21 (*summer solstice*) it is +23.5°. In spring and fall it passes through a midway point ($\delta = 0°$) on March 21 and September 21 (*the equinoxes*).

The declination tells us about the sun's *seasonal* relationship to the earth. To describe the sun's position relative to a given site at a given time of day, we use two angles. The first angle, the *altitude* angle (ALT, α or h) describes the height of the sun in the sky. The second angle, the *azimuth* angle (AZ, a_s, or bearing) describes the compass orientation of the sun. The altitude angle is always measured straight up from the ground. The azimuth angle, as commonly used in architecture, is the sun's position either east or west from due south. For example, if the sun is due south, AZ = 0°; if the sun is due east, AZ = 90° east of south; and if the sun is due west, AZ = 90° west of south. Another convention measures from due north in a clockwise manner. Thus due east results in AZ = 90°, south results in AZ = 180°, and west results in AZ = 270°. Computer programs, even in architecture, tend to use this convention.

SUMMER
LEANING 23.5° TOWARDS THE SUN
$\delta = +23.5°$

SPRING

SUN

FALL LEANING SIDEWAYS TO THE SUN
$\delta = 0$

WINTER
LEANING 23.5° AWAY FROM THE SUN
$\delta = -23.5°$

SEASONS IN THE NORTHERN HEMISPHERE CAUSED BY CHANGING DECLINATION

Consider the sun's behavior in the Northern Hemisphere. In the winter the sun rises south of due east, arcs low through the sky and sets south of due west. In the summer it rises earlier and to the north of due east, arcs high through the sky and sets later, north of due west. At noon the sun is always due south, no matter what the season (north of the tropics). If you notice that sometimes this is not the case, it is because of the difference between legislated time and *sidereal time* (real or solar time). They are the same only at the center of a time zone, but elsewhere, sideral time is either earlier or later than legislated time. In addition, we have daylight-saving time for six months of every year, which advances the "legal" time one hour.

The first conclusion we can draw from the seasonal variation in the sun's path is that there is a lot of sun on south-facing facades during the winter, and not much sun anywhere else. In the summer, on the other hand, there's a great deal of sun on the east facade, the roof, *horizontal skylights*, and on the west facade. This means that southern exposures should be maximized for winter sun, especially if there is an overhang, while eastern and western exposures and horizontal skylights should probably be minimized to reduce summer sun.

Indeed, in temperate and cold climates, this is the basic passive solar design strategy. We want maximum glass on the south facade, but the sunlight that comes in through the glass in the winter should be stored without changing the temperature too rapidly, which would cause overheating. For this reason, there should be sufficient thermal mass, which could be provided in the form of a concrete floor with a conductive finish such as slate or terrazzo. Mass could also be provided by a concrete block wall or other heavy building material. Special solar design prototypes will be discussed later.

Overhangs

South-facing windows should be protected by an overhang which allows the low winter sun to pass under the overhang and in through the window, while the high summer sun is cut off. Thus, the perfect overhang is one which is calculated to admit the winter sun and block the summer sun. This can be done by placing the overhang higher up the wall (at a greater

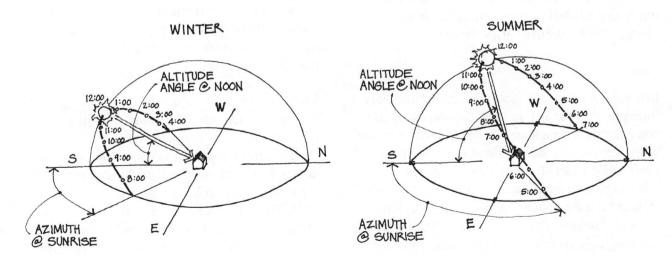

SUN'S PATH ACROSS THE SKY AT 40° LATITUDE

OVERHANG AND PROFILE ANGLE

distance above the window head) and increasing the projection proportionally, or by tilting the overhang to match the incoming winter ALT angle, as shown above.

The angle of the *shadow* line is called the *profile angle* (Φ) and coincides exactly with the altitude angle when the sun is directly facing the wall (perpendicular azimuth). At other times of the day, the profile angle is determined by the interrelationship between azimuth and altitude angle, which varies by season.

Fins

In temperate or hot climates, we can afford very little glass on the east or west facade, especially since the west facade gets heat in the afternoon, when the outside air temperature is highest. *Horizontal* glass should be minimized as well. Clerestories or even lanterns or sawtooth roofs should be used instead of horizontal skylights. In cold climates, the sawtooth can face south (although we have to be careful about glare on the space below), and in warm climates we face the sawtooth north.

VERTICAL FINS ON WEST FACADE

East and *west* facades should be protected with *vertical* fins or a combination of horizontal and vertical fins. If we turn the fins to point slightly south, then the winter sun will be admitted (the sun sets south of west in winter) but the summer sun will be greatly reduced (the sun sets north of west in summer).

Solar Plot

The design tools developed to handle all of this are called the *solar plot* and the *shadow mask*. The solar plot is the path of the sun plotted onto a grid (circular or rectangular), and the shadow mask is the representation of the shading devices plotted onto the same grid, so that they may be compared.

The circular graph comes from drawing a hemisphere over the site. The solar plot comes from the sun's path on that hemisphere, and the shadow mask comes from masking all of the angles which are obscured by the shading device (or trees, other buildings, surrounding terrain, etc.). The radial lines are azimuth lines and all of the concentric circles are altitude lines. While we can plot the path of the sun for any given day, typical diagrams plot one day for each month. This results in seven curved lines, one

VERTICAL FINS ARE
MOST EFFECTIVE IN EAST
AND WEST EXPOSURES

each for December and June, and five for months with identical paths (e.g., January and November, February and October, March and September, etc.). Only December and June are unique.

The same information can be drawn on a rectangular graph, in which case the vertical lines are the azimuth values and the horizontal lines are the altitude angles. The shading mask is very much like a wraparound photo of all of the objects which would obscure the sun and cast a shadow on the point being considered.

The shadow mask allows the designer to check what time of day or year the sun will fall on the window, and when it will not. If a tree or a shading device covers a certain set of angles

on the plot, then it would result in the window being shaded for those sun angles. The designer may add a tree, or an overhang or a fin. Conversely, if another building causes too much shadow at the wrong times, he/she may move the window.

Solar Intensities

There is a secondary result of the solar position. When you shine a flashlight against a wall you get a small round spot of high intensity if the flashlight is held perpendicular to the wall. If you aim the flashlight at an angle, you get a larger elliptical area of lower intensity.

Similarly, the sun's intensity per square foot varies depending on the angle between the wall and the *solar vector* (a line drawn to the sun's position). The highest intensity occurs when a hypothetical receiving surface is perpendicular to the solar vector, and that intensity is called the *direct normal intensity* (I_{dn}).

This direct normal intensity varies with the time of day. In the early morning, because the sun is at a shallow angle, the sun has to pass through a lot more atmosphere. This reduces the intensity available. The combined effect of varying direct normal intensity, and of varying angular relationship between wall and sun, results in a very large variation in the *solar intensity* (I_s or S_g) on a given wall orientation for a given time of day. If you remember the discussion of radiant gain (q_r) and solar intensity (S_g), you will remember that S_g varies depending on orientation and time of day. It is the result of the combined effects of the altitude angle, the azimuth, and I_{dn}.

EXTREMELY COLD CLIMATE...
... COMPACT ARCHITECTURAL FORM

CLIMATE

There are many classifications for climate. The best way to study a climate is to plot it on the psychrometric chart. Let us look at one of the four prototypical climates on a psychrometric chart. Although there are numerous climate zones for calculating energy code compliance, discussing four prototypes is sufficient for our purpose.

Cold

The cold prototype is actually rare within the United States; Alaska and the North Central plains states probably qualify. Any time you see a car from Montana, look for a plug sticking out of the grill. It is there so that the engine block can be plugged in overnight to keep it from freezing.

If we plot the daily temperature cycle for an average day of each month in Billings, Montana, we see that most of the year falls to the left of the comfort zone, meaning that it is too cold. However, certain times of day during the summer months of June, July, and August fall withing the comfort zone. If we could capture daytime temperatures and seal the building at night, we would be in great shape. Also note that the daily loop is almost always the same shape and angle. The temperature and the relative humidity are inversely proportional to each other (high temperature results in low RH,low temperature results in high RH). The peak temperature is around 3:00 or 4:00 in the afternoon, and the peak relative humidity is coincident with the minimum temperature, at 4:00 in the morning. It is not that a lot of

COLD CLIMATE PLOT
(BILLINGS, MONTANA)

moisture has been added to the air, but rather that the same moisture content represents a higher percentage of saturation at the lower temperatures. If the air temperature drops below the temperature at which saturation occurs, we have 100% RH, and the moisture will condense out. This either means rain, or sometimes, just dew. Thus this temperature (for a given moisture content) is called the *dew point temperature.*

The best solutions for a cold climate are building forms which minimize the exposed surface area (remember, $q_c = U\,A\,\Delta T$). The objective is to enclose as much volume as possible within the minimum envelope. The best forms tend to be the simplest; for example, the cube or the hemisphere. An igloo is really an ideal form given the building materials and climate. Most vernacular designs in colder climates are cubical buildings, typically two stories with a big sloping roof. Many cultures have farmhouses which include the barn under the house on the ground floor, and the living space on the upper floors, enclosing all the volume in a single skin. The same basic farmhouse form can be found in Switzerland and Nepal. New England homes had very blocky houses with most of the glass on the south facade. Storage rooms, which weren't terribly important, and the kitchen, which created its own heat, ended up on the north. Evergreen trees were planted in a row or in a large group to the north or northwest of the building, to block the prevailing winter winds. A row of trees will shield objects downwind for a distance of 3 to 5 times the height of the row.

The classic Salt Shaker is a house with a two story south facade, and a long sloping roof to the north, resulting in a single story north facade.

SALT SHAKER

If the builder was confident in his ability to build a firesafe chimney, the fireplace was placed in the middle of the building, adding internal thermal mass. If not, it was on the west end. The entry had an internal or external *vestibule* (now known as an *airlock*) which allowed people to enter or leave without losing all of the heat from inside. The house performed extremely well for the building technology available at the time.

Temperate

Most of the United States has a temperate climate. The winters are too cold and the summers are too hot. The climate plots are below the comfort zone in winter, above the comfort zone in summer, and shift through the zone in spring and fall. If we use air movement, solar input to boost MRT, and proper insulation, we can solve both kinds of problems. The best way to do that is to ascertain from the plot what months require what action, and then superimpose those months on a solar path diagram to see what solar angles should be admitted and when shading is necessary.

The resulting building shape tends to be a modified version of the cold climate building. It has been stretched to the east and west, making the south facade longer (remember the discussion of solar paths in winter and summer). There are often porches to the south, or awnings on the windows. This results in an increase in winter solar gain, without any increase in summer solar gain. We sometimes find a large deciduous tree on the south or west, which provides shade in summer and which loses its leaves in winter. To the north we again find evergreens, if there is any northerly winter wind.

TEMPERATE FARM HOUSE

TROPICAL HOUSE

Hot Humid

The climate plot for a hot humid climate, such as that of Houston, is perhaps the most troublesome. It is out of the comfort zone for much of the time simply because of the humidity level. Furthermore, some of the strategies which designers use for other climates don't help here. Until there was air conditioning, insulation was of no help, and we certainly do not want solar gain. The best that could be done was to allow whatever air circulation there was to dissipate heat as rapidly as possible.

The buildings loosen up even further from the temperate prototype. The kitchen detaches, or on more elaborate buildings, there are two kitchens: one for summer and one for winter. Often there are exterior passageways of one sort or another, balconies or breezeways. In really tropical climates, the walls are sometimes omitted, or reduced to lightweight privacy shields. The house is sometimes elevated on poles to allow air flow underneath. The roof is in two separated layers or completely open at the gable so that warm air rises up and out of the building. Some Southern houses actually use this method to empty out the air through the central stairwell. Using convection to suck fresh air through the building in this manner is called *thermosiphoning*.

The brush is often cleared from window areas to allow the best possible ventilation, and palm trees (if available) are planted. The palm is the closest thing to a natural parasol: there is no impedance to breezes underneath, but still some shade when several trees are grouped together.

Hot Arid

A hot arid climate plot typically shows the greatest daily variations. Hot arid climates, such as that of Phoenix, tend to have large diurnal (day to night) temperature swings. This is because places with an arid climate usually have a clear sky. At night, there are large radiation losses to the sky, both from the building and the environment in general. The large temperature

COURTYARD HOUSE

swing gives us a major opportunity, because if either extreme is within the comfort zone, it can be captured and stored by the building, resulting in a pleasant environment.

In addition, we know that heat can be "used up" by the evaporation of water. If water is available, the air is dry enough so that evaporation within the building compound can bring the temperature down significantly.

The vernacular design for such areas takes advantage of those two factors. It is built of high thermal mass materials, such as adobe. The adobe stores the heat from the day for the following night and stores the coolness of the night for the following day.

The form of the house is the ultimate extension of the progression from tight cube to loose elements: it wraps around the outside environment and pulls it in. The Spanish or Moorish atrium or courtyard house contains a piece of the outside environment in its center, which it modifies by evaporating moisture into the shaded area by the use of a fountain or plantings. The house is open to the courtyard, so the cooled and humidified air enters the house. The only windows on the outer walls are small and occur high on the

wall, allowing the heated air from inside to escape, but forming a "pool" of cooler air, which does not rise. There are psychological factors inherent in this kind of building, as well. The sound of the fountain and the quiet courtyard produce a cooling and calming environment, even if they don't bring about much change in temperature.

PASSIVE SOLAR DESIGN PROTOTYPES

Passive solar design uses the sun to heat the building without any moving parts or pumps. In a sense, this definition is accurate but oversimplified. Passive design also includes buildings which do *not* overheat because they are correctly designed. In fact, it should include all buildings, since all buildings (whether "solar" or not) should take proper orientation and materials into account. The strict definition of a *passive system* is one in which the collector device and the storage device are one and the same (i.e., the structure itself). *Active systems* are those in which the collector and storage are separate (i.e., a collector on the roof and a storage tank in the basement). Most buildings are somewhere in between.

The first principle of good passive design is to do whatever the good vernacular design solutions did. This results in appropriate regional architecture, usually responds to the context, and represents an optimum use of form and materials.

Direct Gain Space

There are some special strategies which have been developed, and some generic names have been given to old strategies. For example, a room in which the structure and thermal mass are in direct sunlight is called a *direct gain space*. This usually means that uncovered, high mass floors are placed in the southern rooms, so that they absorb the sunlight directly. Concrete, stone, terrazzo, tile and other materials which conduct and store heat are ideal for this purpose.

Mass Wall

The first extension of this strategy is to place specially thickened walls directly in the sunlight (typically on the south facade). These walls are often behind a large window, or sometimes an entire glass skin, and are generically called *mass walls*. They store the incoming solar energy without increasing temperature rapidly, and release the heat when needed without dropping temperature rapidly. They often store a great deal of heat. Mass walls also delay the arrival of the heat on the inside, and may be tuned in thickness to match the expected temperature cycles. There are two special types of mass walls: Trombe walls and water walls.

Trombe Wall

The most famous type of mass wall is the *Trombe wall*, named after a Frenchman, Felix Trombe, who popularized the technique. The Trombe wall adds a convective loop to the system, by trapping a layer of air between the wall and the external glass skin. When the air heats up, it rises. There is a one-way vent at the top of the wall which lets the warm air into the room. There is another one-way vent at the base of the wall which lets the coldest air in the room, which is along the floor, into the space between the glass and the wall (which, in turn, is heated, repeating the cycle). This is an

TROMBE WALL

example of a type of thermosiphoning. The wall literally stirs the room air when the sun shines and shuts itself off when the sun does not shine.

Water Wall

The second type of mass wall is the *water wall,* which consists of a tank or a collection of large vertical tubes, filled with water and placed next to the window, inside the building. The tubes

and the water may be clear or colored. It is different from other thermal mass materials in that it allows some of the light through the wall into the room. Water has an extremely high specific heat, storing about 5 times as much heat per degree change per pound as concrete does (or about two to three times as much per cubic foot). Thus, it mitigates temperature fluctuation within the room. (See discussion of C_p in Lesson One.)

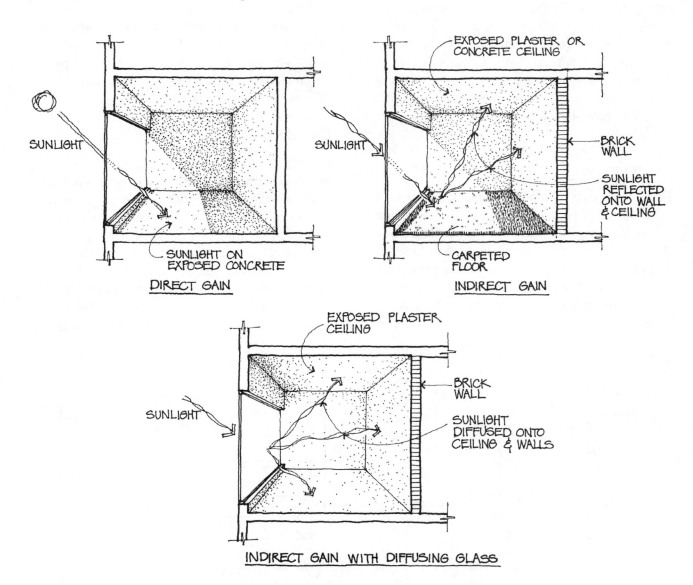

DIRECT GAIN

INDIRECT GAIN

INDIRECT GAIN WITH DIFFUSING GLASS

Indirect Gain

An *indirect gain* space is similar to a direct gain space, in that the usual structural and finish materials have high thermal mass and can absorb and store a great deal of heat without much temperature fluctuation. The difference is that in the indirect gain space, the mass is not directly in the path of the sunlight, but in a shaded area of the room. The thermal mass is heated indirectly, either by reflected sunlight or by the warm air in the room. This strategy can be improved immensely by using a diffusing glass in the window, which redirects the incoming sunlight to hit much of the mass directly. Otherwise, an indirect gain space requires about four times as much thermal mass as a direct gain space, to have the same effect.

Greenhouse

Perhaps the most common simple application of solar design is the addition of a *greenhouse* onto a living space. You remember the definition of the greenhouse effect given in Lesson One. Typically, a fan is connected to a temperature sensing device and moves the air from the greenhouse into the space when the desired temperature differentials are attained. There

should also be an exhaust fan for use when the greenhouse overheats.

Super Insulated

Another old technique which has been revived and perfected is the use of very large amounts of insulation, and most importantly, sealing the construction very carefully. Super insulation, consisting of R-20 walls and R-30 roofs, is used; the seams on vapor barriers are carefully taped and gaps between window frames and wall are foam filled. There are no pipes or conduits in exterior walls. The electrical wiring, switches and sockets are typically mounted on the room *surface* of the wall in a conduit or chase rather than penetrating the wall.

Double Envelope

Double envelope construction consists of a building within a building. The outer shell utilizes passive design, with lots of south-facing glass. The inner shell also involves passive design, with a backup heating system. The inner shell is maintained at the intended temperature, and the outer shell provides a mild and protected climate for the inner shell. The east and west facades often consist of a single superinsulated shell. The result is that a

DOUBLE ENVELOPE CONSTRUCTION

north-south cross section looks like one building inside another, while an east-west cross section looks like a single building with two roof layers. It was originally believed that there would be a convective loop around the building when the sun shone on the south side. However, experiments indicate that some stagnation occurs and that this effect is not as predominant as originally surmised. There does seem to be a connection between the space between the inner and outer shells and the temperature of the ground under the building, resulting in fairly good performance anyway.

Earth Sheltered

Perhaps the most interesting development is the trend toward "earth sheltered" buildings, which may be partially sunken, "snuggled into the hillside," bermed, or totally underground. The greatest thermal benefit in all of these cases is *not* the *insulation* provided by the earth, but rather its tremendous *thermal mass*, with the resulting storage and averaging effects. If you dig deeply enough into the ground, you will come to a layer which never experiences freezing temperatures (the frost line). In fact, we can draw temperature profiles, in which the surface of the ground may be at air temperature,

but just a few inches lower, the temperature is much milder, and at a depth of 20 feet, there is almost no variation from the year round average. Year round average temperatures are much easier to deal with than the extreme air temperature swings we get in most climates. With earth sheltered buildings, there are increased structural costs, and waterproofing is critical, but there are a number of benefits as well: increased security, durability, and privacy, and decreased maintenance, since no short term painting and reroofing are necessary. In addition, more ground surface becomes usable as outside space.

Other Passive Techniques

There are other techniques which are more or less "passive" in nature. In climates with high daytime temperatures and low nighttime temperatures, ventilating the building at night to cool it and closing it during the day can keep the building as much as 20° cooler than the daytime peak. This is called *nighttime flushing*.

The *roof pond* uses sliding insulation panels over a pond or bag of water on the roof. During the summer daytime, the panels are closed and the water absorbs heat radiated from the house below. At night, the panels are opened and the

SLIDING INSULATION

ROOF POND

WINTER DAY
SUMMER NIGHT

WINTER NIGHT
SUMMER DAY

ROOF POND

heat is reradiated to the nighttime sky. This requires comparatively clear climates, since cloud cover reflects ground radiation back down. During the winter, the system may be used in reverse if needed. The panels are left open all day long, exposing the water to the sun, and closed at night to insulate the heat that was gained. This only works, again, where the air is clear and the temperatures mild enough so that the water heats up. In such instances, the water is most often placed in large, black plastic bags, increasing both absorptivity and emissivity, and the insulation panels are foil backed on both surfaces. Radiation is the key thermal process employed.

In climates with snowy winters and hot, humid summers, there have been buildings built in conjunction with an insulated pond, often covered by an insulated dome. All winter long, a snow machine sprays a fine mist of water into the pond, while the dome is opened to the outside air. An enormous pile of heavy snow is formed where the pond used to be. In the spring, the dome is closed off, and the snow/ice begins to melt. The resulting water provides an excellent source of chilled water for air conditioning systems. These projects are economically feasible primarily for rural or suburban campuses, where there is sufficient low cost land for the ice pond.

Energy from the sun or the wind, or even from burning wood, is know as energy from *renewable sources*. Such energy may be utilized for either passive or active systems. Let us look at some of the active systems.

ACTIVE SOLAR SYSTEMS

Active solar systems are typically used to do one of four things: heat water, heat the building, cool the building, or generate electricity. All active systems have some elements in common. These include flat plate or focusing collectors, and rock bed, fluid container, or battery storage devices. In some cases, the existing utility network is actually used as the storage device.

A *flat plate collector* is exactly what the name implies, a flat surface tilted at approximately the right ALT and AZ angles to receive most of the sun's rays as directly as possible. Such collectors usually continue to function even when the sun is not exactly perpendicular to the surface, and also on overcast days when the solar rays are scattered and not parallel.

A *focusing* collector consists of either a parabolic trough or a parabolic dish, or an arrangement of lenses, which focuses incoming light onto a tube or a point. It generates much higher energy densities and much higher temperatures than a flat plate collector. However, it also has to be aimed much more carefully. If the collector is not properly aligned with the sun, then the sun's rays focus on the wrong point altogether. If the sky is cloudy and the solar rays are not parallel, then focusing collectors often don't work very well no matter where they are pointed. This means that most focusing collectors are mounted on gimbals or motorized devices which allow them to be constantly adjusted throughout the day. They are most effective in clear climates.

Domestic Hot Water

The most effective use of active solar systems at this time is to heat water to be used in place of domestic hot water or hot water used in industrial processes. There are several different types of systems, which may be either open or closed loop.

A flat plate collector system uses a collector plate composed of tubes which either have flat fins or are attached to a large metal

SUNLIGHT

COVER PLATE

COLLECTOR SURFACE

FLUID DUCTS

INSULATION

FLAT PLATE COLLECTOR

collector surface. The surface is insulated heavily on the back and sides, and has a cover plate, typically of a very high transmissivity glass or plastic on the front.

A focusing collector system utilizes a parabolic trough with an interior surface of reflective aluminized mylar and a tube, clear or black, running through the focal line of the trough. The clear tubes are usually the size of a fluorescent tube and filled with a colored collector fluid.

PARALLEL SOLAR RAYS

COLLECTOR FLUID TUBE AT PARABOLA FOCUS

REFLECTING SURFACE IN PARABOLIC CROSS SECTION

FOCUSING COLLECTOR

The black tubes are copper or steel pipe with the water running through them.

Some systems use a bent *Fresnel lens* to focus the incoming sunlight. A Fresnel lens uses less material than a normal lens. The surfaces still have the same orientation and curvature, only they are stepped, rather than continuous. The most common Fresnel lens is an automobile headlight, the inside of which is ridged to help focus and direct the light without using a thick lens in front of each lamp.

NORMAL LENS   FRESNEL LENS

FRESNEL LENS

Open loop simply means that the fluid going through the loop is the fluid that will be consumed. For example, the water in the loop will be used for domestic purposes such as cooking or washing.

Closed loop systems use one medium in the collector, which then runs through a water storage tank without being mixed with the water. Usually antifreeze (glycol) is run through the collector and then through a coil inside the water tank. This solves the problem of what to do with the fluid inside the collector overnight. If water were used inside the collector in a cold climate, it could freeze and burst all of the tubes in the collector as well as the connecting piping.

Drain down and *drain back* systems solve the freezing problem by emptying the collector when the temperature drops too low. The drain down system senses the temperature differential and opens valves to drain the fluid down into a reservoir. The drain back system is a failsafe system in which the collector is only full while the pump is running. If the pump shuts off, everything in the collector drains back. The pump is set so that it only turns on when the temperature in the collector is higher than in the storage.

A *batch system* is nearly a passive system, and is simply a storage tank exposed to the sun. In many cases the tank is encased in an insulating box with a folding insulating panel over the glass face of the box. For this reason batch systems have sometimes been nicknamed *"bread-box"* systems.

A *thermosiphon system* is one in which the storage tank is higher than the collector and often immediately adjacent to it. The water circulates between the collector and the tank by convection rather than being pumped. The warmed water in the collector rises into the top of the tank. The coldest water, at the bottom of the tank, is siphoned off to the base of the collector. If the tank is outside the house, it must be well insulated.

Space Heating

Active solar collectors are used for space heating in two ways. A simple and safe way is actually halfway between an active and a passive system. This system is known as *air and rock bed storage*. The collector may be a large flat plate with air ducts rather than water ducts, or even a greenhouse type of space. The heated air is blown through ducting to a large bin full of coarse gravel, often underneath the house.

The air passes between the chunks of gravel, warming them and storing the heat. Later, when the sun has set, the air may be blown through the rock bed in reverse, and then ducted to whatever portion of the house requires heating at the time.

The other method uses water as the medium and is very similar to heating the water for domestic hot water use. The only difference is that the water is piped either to a heat exchanger at the furnace, or more often, directly around the building to fan coil units, or to radiant surfaces or registers, or to baseboard heaters.

Absorption and Desiccant Cooling

It is not surprising that cooling with solar energy is much more difficult than heating. Two methods have been developed. For small scale applications, the *desiccant system* is being tested. This system uses the sun to bake all of the moisture out of a desiccant. Outside air is then brought in past the desiccant, which absorbs all of the moisture. The dried air may be passed into the building, or water may be sprayed through it, the evaporation causing the temperature to drop. The system requires two batches of desiccant, one which is being dried while the other is being used.

The *absorption refrigeration* cycle is a fluid version of the same principle. The sun is used to evaporate moisture out of a brine solution, typically *lithium bromide*, until it reaches saturation. The solution is then used to absorb water *vapor* from a second source of clear water, increasing the evaporation rate in that source, and thus *cooling* it. Water from that second source may then be used for cooling the building. More often, a second water line is run through a heat exchanger in the cooled pool.

WIND

WIND

PLAN

ELEVATION

SAVONIUS

DARRIEUS

VERTICAL AXIS WIND TURBINES

Steam Generation

Solar heat in focusing collectors can not only warm water but can actually generate steam. The primary use for such installations is to generate electricity. On a small scale, this is quite safe, but not terribly efficient.

Photovoltaics

The most efficient method, with a nearly unlimited potential, is the direct generation of electricity from sunlight. This is called *photovoltaic* conversion, or more commonly, *solar cells*. Flat, very thin cells of a semiconductor made from silicon (sand) create an electrical charge (a difference in electrical potential) when they are exposed to light, which can be utilized simply by tapping opposite surfaces of the cell with small wires. This is equivalent to a direct current (DC) battery and can run

lights and other simple electrical devices, or the electricity can be converted to alternating current (AC) and used with common house hold devices or sold back to the utility company. The most effective type of DC to AC conversion is called a synchronous inverter. Photovoltaic cells have been in use for some time, and are already cost effective where the alternative is to put up new power poles and run long power lines. The cost of photovoltaics is expected to continue dropping, much like the cost of calculators and other semiconductor devices over recent years.

The early cells were made of crystalline silicon, first cut from an ingot and later extruded in a ribbon. The most common cheap cell now is an *amorphous* silicon cell, often found on "solar" calculators. There is also some promise shown by gallium arsenide based cells.

WIND ➡

LEADING

WIND ➡

30 FEET ABOVE SURROUNDING OBJECTS

TRAILING

HORIZONTAL AXIS WIND TURBINES

Typical conversion efficiencies range from 10% to 13%. Efficiencies as high as 60% have been attained under laboratory conditions.

Wind Turbines

Another form of technology which has been with us for some time is the windmill. Wind turbines capable of providing for all the needs of a household can be purchased for under $20,000. They generate electricity through a generator or alternator, which can be used directly, or converted to AC and sold back to the utility. Large "windfarms" in California, typically in a mountain pass, are now covered with turbines, and in New England, some consumers generate their own electricity using individual wind turbines. Any area with an average windspeed of 10 mph or more is a good candidate for a wind turbine, and a 13 mph average results in a steady profit from selling electricity to the utility company.

There are two basic types of wind turbine: the *vertical axis wind turbine (VAWT)* and the *horizontal axis wind turbine (HAWT)*. The VAWT is usually a Savonius or a Darrieus turbine. The *Savonius* is composed of two offset cups which spill into each other, much like two halves of a steel drum which has been sliced vertically and slid halfway apart. The *Darrieus* resembles a single eggbeater stuck in the ground. Both turbines work regardless of the direction of the wind. The Savonius is not as efficient, but is self starting. The Darrieus is very efficient, but will not start itself. However, it picks up speed nicely, once it is going.

The horizontal axis turbine is the more common type today, and is available in leading or trailing blade configuration. If the blades are upwind of the tower, it is called the leading configuration; if they are downwind of the tower, it is called trailing. Keeping the blades upwind (leading) requires either a tail, like a weather vane, or some arrangement of gears and a motor which is attached to a smaller wind direction sensing device. The downwind (trailing) configuration results in eccentric vibration, caused by the temporary release of pressure on the blade as it passes through the downwind "wind shadow" of the tower.

PRINCIPLES OF SUSTAINABLE DESIGN

Why is sustainable design necessary?

Principles of the Scientific Laws of Nature.

1. In the earth's ecosystem (the area of the earth's crust and atmosphere approximately five miles high and five miles deep) there is a finite amount of natural resources. Mankind has become dependent on elements such as fresh water, timber, plants, soil, and ore, which we process into necessary pieces of our human environment.

2. Given the laws of thermodynamics, matter cannot be created or destroyed. The resources that we have been allotted to manage our existence are contained in our ecosystem.

3. All forms of energy tends to seek equilibrium and therefore disperse. For example, water falls from the sky, settles on plants, and then percolates into the soil to reach the subterranean aquifer. Toxic liquids, released by humans, and exposed to the soil, will equally disperse and eventually reach the same underground reservoir. The fresh water aquifer, now contaminated, is no longer a useful natural resource.

These laws of science indicate why it is necessary to maintain the delicate balance of natural ecosystems. There is a need to focus on the preservation of beneficial natural elements and diminish or extinguish natural resources contaminated with toxins and our destructive practices.

There are many credos for environmental responsibility. One, *The Natural Step*, was organized by scientists, designers, and environmentalists in 1996.

They were concerned with the preservation of the thin layer that supports human life in a small zone on the earth's surface: the ecosphere (five miles of the earth's crust) and the biosphere (five miles into the troposphere of the atmosphere).

Their principals are as summarized as follows:

1. Substance from the earth's crust must not systemically increase in the ecosphere. Elements from the earth such as fossil fuel, ores, timber, etc., must not be extracted from the earth at a greater rate than they can be replenished.

2. Substances that are man-made must not systemically increase in the ecosphere. Man-made materials cannot be produced at a faster rate than they can be integrated back into nature.

3. The productivity and diversity of nature must not be systemically diminished. This means that we must protect and preserve the variety of living organisms that now exist.

4. In recognition of the first three conditions, there must be a fair and efficient use of resources to meet human needs. This means that human needs must be met in the most environmentally sensitive way possible.

Buildings consume at least 40 percent of the world's energy. Thus they account for about a third of the world's emissions of heat-trapping carbon dioxide from fossil fuel burning, and two-fifths of acid rain-causing carbon dioxide and nitrogen oxides.

Source: David Malin Roodman and Nicholas Lessen, "Building Revolution: How Ecology and Health Concerns are Transforming Construction." Worldwatch Paper 124 (Washington DC, Worldwatch Institute, 1995).

Page 10, The HOK Guidebook to Sustainable Design, Sandra Mendler & William Odell, John Wiley & Sons, Inc., New York, 2000.

The built environment has a monumental impact on the use of materials and fuels to create shelter for human beings. The decisions about the amount and type of materials and systems that are employed in the building process have an enormous impact on the future use of natural resources. Architects can affect and guide those decisions of design to influence the needs of sustainability and environmental sensitivity.

It is not a quick process. Like moving a large boat in a new direction, it must be done gradually and with awareness of the many natural forces that are acting on it. But, the process has started and architects should be aware of the philosophy of sustainable design in order to influence the results.

COMMUNITY SYSTEMS AND ZONING

There are several ways in which communities may become involved in solar energy. A utility or syndicate may build a renewable energy power plant, or a municipality may enact zoning laws which specify solar access rules for a neighborhood, or even mandate the installation of solar domestic hot water heaters on all new residential construction.

Power towers are stationary towers with a collection of solar panels or a tank on top. A surrounding field of rotating mirrors aims the sun's reflection onto the tower, immediately turning the water to steam, which in turn is piped down the tower and used to generate electricity. One such installation was *Solar One*, in Southern California, which produced 10 megawatts of power using a 72 acre field of mirrors.

The Luz Company continues to operate fields of parabolic trough focusing solar collectors generating electricity at a profit.

In large building complexes, where there is a central steam plant to heat the entire facility, a technique called *cogeneration* is employed. Electricity is generated on site by heating water to steam and driving steam turbine generators. The byproduct steam is then used for heating the buildings. This is much more efficient than simply heating the steam and pumping it around to heat the buildings. In some communities, garbage is burned and the waste heat is used to generate steam and thus electricity.

Community zoning laws can create *solar envelopes* for a site, similar to other zoning restrictions, such as setbacks and floor area ratios. A solar envelope is defined by requiring that nothing built on the site may cast a shadow off the site, or in some cases cannot entirely shade anything 10 feet tall which is off the site. In effect, this protects the right of everyone to enjoy sunlight.

THE SUSTAINABLE DESIGN PROCESS

Is a sustainable design organized and implemented differently than a conventional design?

The Design Team

What kind of design team is necessary for a sustainable project?

The scope of a sustainable design invites an expanded team approach, which may include the following:

■ Architects or engineers (structural, MEP) with energy modeling experience.

- A landscape architect with a specialty in native plant material.
- A commissioning expert (if LEED employed).
- An engineer/architect with building modeling experience.

The design team for a sustainably designed project tends to have a larger pool of talent than a typical architectural project. Because the buildings will be more holistic, the sustainable design team will have additional consultants that bring a broader range of experience and innovation to the project. Wetlands scientists, energy efficient lighting consultants, native plant experts, or commissioning engineers are examples of the additional talent that may be added to sustainable design projects.

As with any architectural design, there is a hierarchy of design goals:

- *Initial imperatives* such as budget, timing, image, and program necessities.
- *Subjective goals* such as a functionally improved and more pleasing work environment, pleasing color schemes, and landscaping that complements the architecture.
- *Specific goals* such as more open space, more natural light, less water usage, and adjacency to public transportation.

And with the inclusion of sustainability there may be additional goals:

- *Initiatives that are specific to sustainability* such as fewer toxins brought into the space, daylighting in all spaces with people occupancies, less overall energy consumed, less water usage, adjacency to public transportation, and improved indoor air quality.
- Desire to exceed existing standards such as ASHRAE, USGBC, or American Planning Association (APA).

SUMMARY

We have discussed what conditions are comfortable thermally and some tools used to study and express different climates and human comfort ranges. We looked at four climates and at the shape traditional buildings take in response to those climates, and why. We reviewed the basics of solar position and orientation. Using that information, and what we knew of basic heat transfer methods, we examined different prototypical solar design strategies and the terms associated with them.

LESSON 2 QUIZ

1. If the MRT (mean radiant temperature) is low, which way does the comfort zone shift?

   I. Toward higher ambient temperatures

   II. Toward lower ambient temperatures

   III. Toward lower relative humidities

   A. I only
   C. III only

   B. II only
   D. II and III

2. What is the definition of relative humidity?

   A. The amount of water in the air in lbs. of water per 1,000 cu.ft. of air

   B. The amount of water in the air in lbs. of water per pound of air

   C. The amount of water in the air in grains of water per pound of air

   D. The percentage of saturation at the current temperature

3. What is a sling psychrometer?

   A. A chart of the human comfort zone

   B. A chart of air temperatures and humidities

   C. A method of measuring wet bulb temperatures with a manual device

   D. A method of measuring the psychological effect of diurnal temperature variation.

4. What is a Trombe wall?

   A. An unusually heavy mass wall

   B. A mass wall which uses water as the storage element

   C. A mass wall which causes a convective loop into the room behind it when it is warmed

   D. A mass wall that stores energy using phase change materials

5. A good design for a hot arid climate uses which of the following?

   I. Thermal mass to damp out temperature swings

   II. Large openings to the outside to improve ventilation

   III. Evaporation to increase humidity

   A. I, II and III
   C. II and III

   B. I and II
   D. I and III

6. The time of day affects which of the following factors?

   I. The declination angle

   II. The altitude angle

   III. The azimuth angle

   IV. The intensity of the sun

   A. I and II
   C. II, III and IV

   B. I and III
   D. I, II, III and IV

7. Given the psychrometric chart on page 25, what is the wet bulb temperature if the dry bulb temperature is 60°F and the relative humidity is 70%?

   A. 50°F
   C. 96°F

   B. 60°F
   D. 104°F

8. Which of the following would NOT cause a change in enthalpy?

   A. An increase in the velocity of the air movement

   B. A decrease in relative humidity

   C. An increase in the dry bulb temperature

   D. An increase in the absolute moisture content

9. The major benefit to earth sheltered housing is derived from which of the following factors?

   A. The thermal storage capacity of the earth

   B. The insulative capacity of the earth

   C. The available moisture in the earth

   D. The weight of the earth

10. Which of the following statements is NOT true?

    A. Flat plate collectors will continue to function even when they are not pointed exactly at the sun.

    B. Focusing collectors can achieve higher temperatures than flat plate collectors.

    C. Flat plate collectors can be used for heating water or air.

    D. Focusing collectors are easier to mount than flat plate collectors.

MECHANICAL EQUIPMENT AND ENERGY CODES

INTRODUCTION

In the previous lessons we discussed how to reduce heating and cooling loads on buildings by being responsive to the climate and by using solar design, and we also described how to calculate thermal loads. In this lesson we will discuss the concepts and the mechanical equipment for heating, ventilating and air conditioning (HVAC) a building in response to those loads.

Generally, this equipment consists of two basic elements. The first, called the *plant*, creates the warm or cool water or air, usually in the mechanical equipment room. The second is the *distribution* mechanism or *system*, which then delivers that heated or cooled air or water to the necessary rooms or areas in the building, called *zones*.

PLANT TYPES

The scale of a plant may vary, from a room air conditioner, through the equipment in a mechanical room in a building, to a steam plant used to heat an entire campus. Various plants may be used with various distribution systems.

Boilers and Chillers

The early plants were primarily for heating only and consisted of sources of hot water or steam. As air conditioning developed, plants became sources of chilled water or cooled air as well.

Water was heated by lighting a fire under a tank or heat exchanger tube, called a *boiler*. There had to be a separate exhaust *flue* to vent the byproducts of combustion from the building.

REFRIGERATION CYCLE

Today, in energy efficient designs, the air used for combustion is sometimes brought in directly from outside, instead of wasting the warmed and humidified building air. This is called *external combustion air*.

The modern *forced air furnace* duplicates the boiler in residential applications. Air from within the home is brought through a manifold inside a larger combustion chamber. Oil, natural gas, or propane is burned inside the chamber, warming the manifold, which warms the air inside. The combustion air is vented through a flue. Historically, convection moved the warmed *supply air* from the manifold up into the residence (this was called *gravity* feed) but this arrangement required the furnace to be located in the basement, and also didn't move the air fast enough. As a result, a fan was introduced, forcing the air returned from the home (*return air*) through the manifold and into the ducts no matter what the height relationships were. If the flow is downward through the furnace (reversing the convection) it is called a *downdraft* furnace. If the furnace is reduced to approximately five feet in height so that it can fit in an overgrown closet or even an attic it is called a *lowboy*.

Refrigeration Cycle

Modern air conditioning relies primarily on the refrigeration cycle. The refrigeration cycle uses a special fluid, *Freon* (which is actually a family of several chlorofluorocarbon or CFC gases), circulated in a closed loop. The *pressure* in the loop is varied using a pump and a constricted section of tubing or a valve, causing changes in temperature and evaporation and condensation.

The pump increases the pressure of the fluid, forcing the Freon to condense, thus releasing the latent heat of evaporation. This part of the loop is called the *condenser*. After the Freon passes through the condenser, it passes through an expansion valve, or simply a constriction in the tube, which results in a pressure drop on the downstream side. This drop in the pressure allows the liquid to evaporate, absorbing the latent heat of evaporation from its surroundings. This is called the *evaporator*. Because of the extreme pressure differences, the condensation goes on at a very high temperature and loses heat to its surroundings. The evaporation goes on at a very low temperature, and absorbs heat from its surroundings. This means we can move heat from a lower temperature to a

CHILLED WATER OR AIR TO BUILDING

HEATED WATER

COOL

HOT

SPRAY

EVAPORATIVE CHILLER

FAN

BLOWDOWN

REFRIGERATION IN BUILDINGS

higher temperature, which would otherwise be impossible.

Both the evaporator and the condenser are usually heat exchanger coils which heat or chill another fluid. On the condenser side the coil often transfers this heat into water being circulated through an *evaporative chiller* outside the building which dissipates the heat into the surrounding outside air. This is especially likely with larger buildings or building complexes. The evaporative chiller is often called a *cooling tower*. It is that large box with louvers exhausting humid air or even mist which you so often find next to an office building. Some cooling towers may function for several buildings. There is a constant loss of water due to evaporation, so water must be added. Unfortunately, this means that dirt and minerals are also added, and left behind when the water evaporates. These must be drained off through a small valve in the base of the cooling tower called a *blowdown.*

On the evaporator side of the cycle, the coil takes the heat from water or air which is typically brought down to 50–55°F and then circulated around the building. This is the cooling loop for the building. In the seasons when it is cool enough outside, the outside air

may be used directly and the refrigeration cycle shut off. In some climates, cool water from a clean pond may be used in early winter and late spring, instead of chilled refrigerated water, or this pond water may be use for cooling the condenser without using an evaporative chiller. Such seasonal adjustments in the source are often referred to as an *economizer cycle.*

Heat Pump

Now, what would happen if we ran the entire system in reverse? That is to say, what if we refrigerated the outside, and collected the heat inside the building? Note that the refrigeration cycle does not create heat or make it disappear, but rather just moves it around. However, it does so very effectively. In expending 1 Btu of energy, we can actually move 2 to 4 Btu's. Normal boilers and furnaces run at about 80% efficiency. If we use a refrigeration cycle to heat a building, we can get the equivalent of 300% because we have *moved* heat energy from the outside to the inside of the building, in addition to the energy expended. We are not really creating energy, so we cannot call this the efficiency, but rather the *coefficient of performance (COP).* The definition of the two are basically identical. The difference is that the COP includes the heat delivered from the outside.

Efficiency = energy delivered / energy used.

COP = energy delivered / energy used.

Typical COP's vary between 2 and 3, with 2.3 being common. If the compressor is placed inside the building, much of the energy dissipated in friction, etc. is also gained as heat for the building.

There is some concern that all of the Freon (or any CFC) used in a refrigeration cycle will eventually leak into the atmosphere. It has been found that CFC's destroy ozone at very high altitudes. Enough ozone has already been destroyed so that there is a hole in the ozone layer over the South Pole which appears every spring. Several chemical companies are now working on alternative gases (typically hydrofluorocarbons or HFC's) for use in the refrigeration cycle. Over the long term, this is absolutely essential.

SYSTEM DISTRIBUTION TYPES

There is much more variation in distribution systems than in plant systems. However, there are only three basic categories: electrical, hydronic, and forced air.

Electrical Systems

Electrical systems are the simplest and the lowest in first cost, and generally the most expensive in life cycle costs. They are justified only in very mild climates where the system is off most of the time. There are two categories of electrical systems: radiant systems, which consist of radiant panels or wires embedded in the ceiling, and baseboard heaters, which heat up and cause convective air circulation in a room.

The advantages of the radiant systems are that the system only needs to be turned on in the rooms currently occupied, and that only objects, such as people, are heated, not the air (remember radiant heat transfer). However, since electricity is often generated from combustion at a low efficiency, this is a wasteful and often expensive way to use energy.

Hydronic Systems

Many hydronic systems are also radiant systems. Hot water or steam is circulated through registers or even through pipes embedded in the floor, which then radiate the heat into the space. Baseboard heaters using hot water or steam are also common. Hydronic systems may also be combined with forced air systems. For example, the hydronic system may bring hot and chilled water to each zone, and then an air circulating unit within the zone uses the water to heat or cool the air at that point, which is then blown into the room.

There are several different loop patterns to the piping in hydronic systems.

Single pipe is exactly what the name implies: a single supply and return pipe, which may be run in series or partly parallel. The hot water is circulated through each register (or fan coil unit, etc.) and back to the pipe. The first register is quite hot, and the temperature decreases with each successive register. This system has a low first cost, but may only be extended a very limited distance, since the water would be quite cold by the last register if there were more than five.

A *two pipe* or parallel system uses separate supply and return pipes. The water is not mixed back into the supply pipe after it has been through the register. Thus, the supply water for each register is at the same high temperature, since it hasn't been adulterated by cooler return water.

SINGLE PIPE

TWO PIPE

THREE PIPE

FOUR PIPE

HYDRONIC SYSTEMS

If both heating and cooling are desired, a separate two pipe system may be provided for each, resulting in a *four pipe system*. The temperatures available at each fan coil unit are fairly constant. Four pipe systems are used with two separate heating and cooling registers, rather than by mixing hot and cold water in a single register. The return pipes are kept separate.

A *three pipe* system mixes both the hot and cold water in a common return pipe. This saves on piping costs, but is more expensive to operate because all of the water must be heated or cooled from the middle temperature instead of having relatively hot and cold water in the separate return pipes.

Forced Air Systems

Forced air systems distribute heated or cooled air around the building using *supply ducts*. The air from the spaces may be returned through return ducts, or may be returned using the space between the suspended ceiling and the floor or roof above as one huge slow-moving air duct, called a *plenum*. In that case, the supply ducts may also be inside the interstitial space between ceiling and the floor above, but the supply air is sealed within the supply ducts and therefore does not mix with the return air. In some instances, the occupied spaces are used as the return plenum, through which the air slowly moves back toward the equipment room.

Doors have registers in them, and sometimes there is a *cold air register* between two floors, so named because return air to a residential furnace is cooler than the supply air. These registers allow the return air to move from space to space while maintaining visual privacy.

The air brought into the plant is usually a mixture of air returned from the building and fresh air brought in from outside the building. This means that there must be a *fresh air intake*, preferably away from the cooling tower and other exhausts, and a duct from the intake to the mechanical equipment room. The outside air must be warmed or cooled, and the return air must be cleaned or filtered. The number of air changes in a room is specified by code or by good practice, and must be provided even if that volume of air is not necessary for meeting the thermal load.

The air is pushed from the plant out through the ductwork by the supply fan, which must have sufficient flow volume to provide enough air and sufficient pressure to overcome the friction in the duct. This pressure is usually expressed in terms of static head of water, which is the height of a column of water which could be lifted by the pressure. (See also Lesson Seven.)

If we want to eliminate infiltration in a building, we positively pressurize the building, by running the supply fan at a rate greater than the sum of the return air fan rate and the leakage rate of the building.

The temperature of the air as it leaves the equipment room is called the equipment temperature or the *deck temperature*. For example, the *cold deck* might be 55°F. As the air moves through the supply ducts, it tends to warm up somewhat before arriving at the room. Insulating the supply ducts keeps this to a minimum. In a supply duct, exterior insulation is used for

temperature control, while interior insulation is usually for sound control.

Fans must be isolated from the floor on which they sit and from the ducts to which they are connected so that their vibrations are not transmitted as noise throughout the building. This is usually done by mounting each fan on springs, which are mounted on rubber pads, with the entire assembly placed on a massive concrete pad. The duct connections are rubber or even fabric, so that the pressure and flow are maintained, but the connection is flexible.

The simplest forced air system is the *single duct* system, often found in residential applications. The amount of air sent out through the ducts is constant, which is called *constant volume*. The furnace simply runs until the temperature reaches the thermostat setting. It is impossible to heat one room and cool another simultaneously. All rooms may be heated, or all rooms may be cooled if there is a central air conditioning unit. The rate of heating may be varied from room to room by putting a *damper* on each *diffuser*, which is simply a metal flap which shuts off or reduces the airflow coming through the louvers or grill at the room. If the air needed for ventilation is greater than the air needed for heating, there is a problem.

A variation of this system is the electric reheat system, in which all of the air is cooled, and an electric resistance heater is placed in the duct upstream of each diffuser. In rooms which need to be heated, the cooled air is simply reheated by the coil. Needless to say, this is not terribly efficient, and is appropriate only for climates where heating is rarely necessary.

The *double duct*, or *dual duct*, system is a combination of two single duct systems, one carrying heated air and one carrying cooled air. This requires twice as much duct space, but

SINGLE DUCT CONSTANT VOLUME SYSTEM

can heat one room and simultaneously cool another. The airflow rate in the ducts is constant, but the amount pulled from each duct at each room is controlled by dampers, and mixed in a *mixing box* at a ratio controlled by the thermostat to provide the appropriate temperature. This system requires more space than any other system, but it is ideal for linear buildings with many different thermal conditions. The two ducts

are usually run parallel above a central corridor, with each zone connected to both ducts.

In order to reduce the amount of space taken up by ducting and the cost of the duct work, the *multizone* system was developed. It is similar to the double duct system, only the mixing boxes are inside the mechanical equipment room, and pre-mixed air is sent out to each zone. If the

DOUBLE OR DUAL DUCT SYSTEM

MULTIZONE SYSTEM

building is square in plan, and there are few zones, this system is efficient. However, if the building has many zones, then many small ducts must be sent out rather than two large ducts, and the system becomes much less economical.

The *fan coil* system is one of the most efficient systems which can heat and cool simultaneously. A constant volume of cleaned and conditioned air is supplied from the plant in a single duct. A chilled water and a hot water

pipe are also supplied. At each zone there is a unit with a fan and two coils. The air is blown over the coils into the room. Hot water may be run in the hot coil if the room needs to be heated, cold water in the cold coil if the room needs to be cooled, and no water at all if the room just needs air for ventilation. Unfortunately, the first cost is rather high, since there is a great deal of plumbing (it is essentially a three or four pipe system) and as much sheet metal as for a single duct system.

FAN COIL SYSTEM

VARIABLE AIR VOLUME

The most common efficient system is the *variable air volume (VAV)* system. This is a single duct system, or there may be three or four separate single duct systems, each of which serves a zone, which may be composed of several rooms. Instead of a constant flow volume, the flow rate may be varied. It is impossible to simultaneously heat one room and cool another within the same zone, but different zones can be handled differently. Thus, one zone may be heated while another is being cooled.

All of the air going to a given zone is at the same temperature, and the *amount* of heating or cooling delivered is determined by the *volume* of air delivered. The temperature of the air in the duct or the overall air flow rate may be constantly adjusted so that the coolest, or hottest, room is just barely taken care of, and thus the system runs at high efficiencies.

The term *unitary systems* covers many kinds of systems. If the air comes directly from the outside through the unit and into the room, then it is a form of unitary system. There is one unit for each zone, often placed on the roof above the zone, or in a permanent cabinet along

a wall. The unit is usually self contained, requiring only electricity, but it may be connected to a chilled or heated water supply which originates elsewhere. Unitary systems are employed most often where the building is spread out and ducting would be impractical or costly because of the length of the run and the temperature drop, and/or lack of space. Unitary systems are also one of the systems used when each zone must have a separate utility bill.

A *heat pump system* is a term applied to a group of heat pumps which serve a building. The system can be quite efficient, although the first cost may be higher. Water is circulated throughout the building and is called the *heat sink*. Each zone has its own heat pump and a fan and short ducts to recirculate air within that zone. The heat pump either removes heat from the water and adds it to air blown into the zone, or removes heat from the zone air, adding it to the water. At certain times of the year, this all balances out, and the water just needs to be circulated. If all the heat pumps are being used for cooling their rooms, the water temperature begins to rise and a chiller in the mechanical equipment room cools it back down. If the water temperature

ZONE 1

ZONE 2

UNITARY SYSTEMS

begins to drop, it is routed through a boiler, which heats it back up.

There are several other terms which are used in forced air systems. *Induction* is the term applied to any system in which a small amount of supply air at a very high velocity is delivered to a box-like unit and mixed with air brought in from the room, which induces a greater airflow than just the supply air delivered to the room.

There are various types of fans. We are most familiar with the bladed fan, similar to a propeller in form and function. When moving

large amounts of air, a *centrifugal fan*, sometimes called a *squirrel cage blower*, is often used.

In addition to heating, cooling or dehumidifying the air, the air must be cleaned prior to circulation or recirculation. *Fibrous filters*, similar to home furnace filters, remove a great deal of the dust and lint, and must be replaced regularly. *Electrostatic filters* are more expensive, but produce less resistance to air movement. They are composed of two sets of charged plates which attract dust particles, which are then cleaned or washed off. *Activated charcoal filters* remove odors and many chemicals from the air,

HEAT PUMP SYSTEM

SUPPLY AIR

ROOM AIR

INDUCTION

SQUIRREL CAGE

INDUCTION UNIT AND SQUIRREL CAGE BLOWER

but are used only when this is necessary. Such filters present a significant resistance to airflow, require very low velocities, and must also be replaced regularly.

PLANT AND DUCT SIZING

Architects and architectural candidates generally do not need to know how to design the mechanical equipment in detail. However, he or she should know how much space it will require. There are two parts to that question: How much floor space will the mechanical equipment in the plant take? And, how much space will the distribution system take in the building cross section?

Plant Sizing

Although it varies with the complexity of the system, the floor space required for the mechanical equipment is usually five to ten percent of the total floor area of the building. This space may be provided in the basement or in a penthouse on the roof. In either case, the architect must provide proper access to the space, since the equipment will eventually wear out and require maintenance or replacement. In a high-rise office tower, one out of every fifteen

or twenty floors may be given over entirely to mechanical equipment.

The capacity of the plant must be sufficient to take care of all of the loads experienced on the design day, based on the efficiency of the distribution system. Heating loads may be expressed in thousands of Btu's per hour or kBtuh. Cooling loads are often expressed in tonnage. A *ton* of cooling is equivalent to *12,000 Btuh*, which is the rate of heat transfer which would melt a ton of ice over a 24 hour period.

System Sizing

As far as the distribution system is concerned, no simple rules of thumb apply. Different systems require different amounts of space, and the layout of the system determines where that space is needed. The forced air systems are the most space consuming. In forced air systems there is a relationship among space, volume, velocity, and noise. Higher flow volumes require either greater cross-sectional area for the duct or higher velocities. Higher velocities produce greater friction, which often creates a noise problem.

The numerical relationship is expressed by the formula:

$$A = 144\ Q_{cfm}/v$$

where

Q_{cfm} = the flow rate in cubic feet per minute (remember that infiltration used Q measured in cubic feet per hour)

v = the velocity measured in feet per minute

A = the cross sectional area of the duct in square inches

Q_{cfm} may be determined from the formula:

$$Q_{cfm} = q_{tot}/1.08 \ (T_{eq} - T_i)$$

where

q_{tot} = the total thermal load in Btuh

T_{eq} = the temperature of the air supplied in the duct (55°F for cooling or 140°F for heating)

T_i = the desired interior temperature for the room

Duct Sizing

We can size the ducts by choosing an appropriate velocity and checking whether the resultant duct size fits into the ceiling cavity, or by choosing the duct size and checking if the resultant velocity will cause excessive noise or too much friction for the fans to overcome.

Appropriate velocities range from 300 fpm for quiet and barely noticeable velocities at the diffuser to 2,000 fpm in a duct. Some large office buildings use vertical ducts or ventilation shafts with velocities in the 10,000 fpm range.

Duct sizes are given in square inches of cross sectional area. For example, a 12 × 12 inch duct is 144 in², and a 20 × 7 inch duct is 140 in². The most efficient duct for a given area is the one with the least perimeter, since that results in the least resistance to air movement and the least friction. For this reason, circular cross sections are the optimum shape, if there is sufficient space. Ducts are often specified in terms of their *equivalent circular diameter* (the circular diameter that would result in the necessary cross sectional area).

Fan Sizing

The above equations can yield the duct size given the necessary flow rate and a desired velocity. There is a another factor which must be considered: the friction of the air traveling through the ducts. For this reason, the duct sizes are often determined from a graph which plots velocity, flow rate, and duct size, in addition to friction loss.

The friction loss is expressed in inches of water per hundred feet, also know as the *static head*. One inch of static head is the pressure required to support a one inch column of water. (Normal atmospheric pressure would support a 33 foot column of water.) This information is used in two ways. First, the rating of the fans in the mechanical room is compared with the friction loss through the system to make sure that they can provide sufficient pressure to overcome that loss and push the air to the farthest diffuser at the flow rate required. Secondly, if the friction loss becomes excessive, larger duct sizes can be chosen, or special fans specified.

ENERGY CODES

Energy codes of one sort or another are becoming increasingly common across the country. There are two basic types: prescriptive codes and performance codes. A *prescriptive code* specifies how to build a building, while a *performance code* states what the final result needs to be and how it will be measured, but does not specify how that result is to be achieved. There are various guidelines, as well. The best known is the *ASHRAE 90-xx* series, which covers suggested practice in the areas of external envelope, HVAC

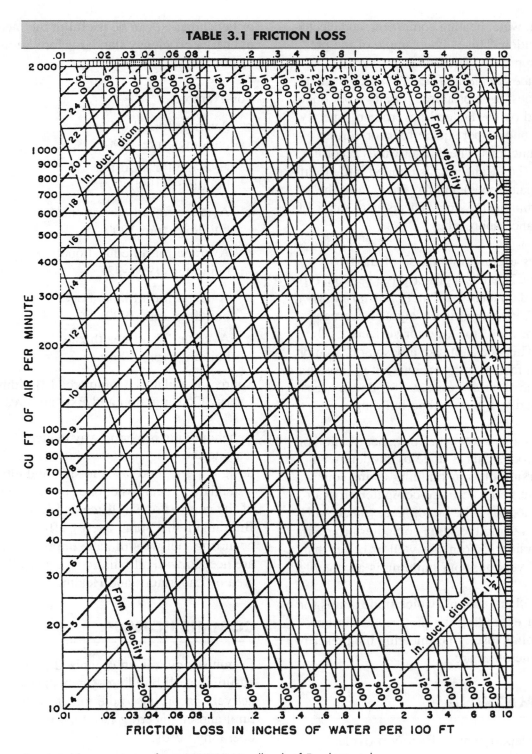

TABLE 3.1 FRICTION LOSS

FRICTION LOSS IN INCHES OF WATER PER 100 FT

CU FT OF AIR PER MINUTE

Reprinted by permission from ASHRAE Handbook of Fundamentals.

equipment, water heating equipment, and electrical distribution. This is an example of the prescriptive approach. The *Building Energy Performance Standards (BEPS),* which have, at various times, been proposed for any work done or financed by the Federal government, specify an energy budget per square foot for various building functions. The budget varies according to climate (i.e., California is divided into 16 different climate zones) and the function of the spaces within the building. This is an example of a performance approach.

Most states use a combination of the two approaches. There is a prescriptive version, typically based on an *overall thermal transmission value (OTTV),* which is a weighted average U-value for all of the exterior surfaces of the building. This does not take any solar design or orientation factors into account, but is based on the "thermos bottle" concept: if it doesn't leak heat too badly, it's not a bad building.

Most states also allow a performance version of the same code. This is done by modeling an OTTV version of your building (same square footage, etc.) on a certified computer program, and then modeling the building as you actually designed it, using the same computer program. If your version performs as well or better than the OTTV version, your building is acceptable.

There are other versions, which trade points for meeting or exceeding the prescription in one area for points lost by not meeting the prescription in some other area.

BENCHMARKING

The U.S. Department of Energy provides "benchmark" information of total energy consumption in BTU's/SF for various kinds of buildings in the United States. These standards,

or benchmarks, can be useful in the measuring of energy efficiency standards for various types of buildings:

For example:

- Average for all office buildings (pre-1990) — 104.2
- Average for all office buildings (1990–1992) — 87.4
- Average for all educational buildings (pre-1900) — 87.2
- Average for all educational buildings (1990–92) — 57.1
- Average for all laboratory buildings (pre-1990) — 319.2
- Average for all health care buildings (pre-1990) — 218.5

Source: U.S. DOE, Commercial Building Energy Consumption and Expenditures.

Benchmarking is a good way to alert the design team to the base energy standards for their design. It's a good place to start and ultimately a standard to beat. And, one can see from some of the comparisons (Office and Educational Buildings), that some energy efficiency is occurring, but we still have improvements that are needed with all buildings, especially in the area of Laboratory and Health.

COMMISSIONING

Commissioning is an organized process to ensure that all building systems perform interactively according to the intent of the architectural and engineering design, and the owner's operating needs.

Commissioning usually includes all HVAC and MEP systems, controls, ductworks and pipe insulation, renewable and alternate technologies, life safety systems, lighting controls and daylighting systems, and any thermal storage systems. Commissioning also verifies the proper operation of architectural elements such as the building envelope, vapor and infiltration control, and gaskets and sealant used to control water infiltration.

Commissioning is a process required for LEED certification, but is a recommended procedure for any building involved with sustainable design procedures.

Source: Commissioning Requirements for LEED Green Building Rating, Version 8. February 5. 1999.

Page 71, The HOK Guidebook to Sustainable Design, Sandra Mendler & William Odell, John Wiley & Sons, Inc., New York, 2000.

INNOVATIVE TECHNOLOGIES

Besides the aforementioned issues of solar design, improved lighting systems, improved HVAC systems, and improved building massing and envelope design, there are several "innovative technologies" that the architect can offer to their project team for consideration.

1. *Ground Water Aquifer Cooling and Heating (AETS)*

   One alternative to full air-conditioning with chillers, which make heavy demands on electricity, is the aquifer thermal energy storage that uses the differential thermal energy in water from an underground well to cool a building during summer and heat a building in the winter.

This is an efficient and relatively low cost system, but it may require approval from the local environmental authority before installation.

2. *Geothermal Energy*

   Where appropriate, heat contained within the earth's surface causes macro-geological events (such as underground geothermal springs or lava formations) that may be tapped to produce heat for adjacent structures.

   In select locations this heat energy can be transferred and conveyed to supplement a building's heating demand.

3. *Wind Turbines*

   Small-scale wind machines used to generate electricity can be mounted on buildings or in open space nearby. These systems have the advantage of being relatively cost effective; a tested and established technology; systematically started-up; and have a relatively high output.

   These systems have the disadvantage of needing a relatively high mast; requiring substantial structural support; potentially causing noise problems; and being visually intrusive.

4. *Photovoltaic Systems*

   The basis of the PV systems is the concept that electricity is produced from solar energy when photons or particles of light are absorbed by semi-conductors.

   Most PV systems are mounted to the building (either on the roof or as shading devices above fenestration). Currently, PV systems are not cost effective. But with promised government subsidy necessary to achieve an economy of scale, PV's may be a viable method of electrical production in the United States, Japan, and Germany in the near future.

5. *Fuel Cells*

Even though Sir William Grove invented the technology for the fuel cell in 1839, it has only recently been recognized as a potential power source for the future. The fuel cell claims to be the bridge between hydrocarbon economy and the hydrogen-based society.

Fuel cells are electrochemical devices that generate direct current (DC) electricity similar to batteries. But, unlike batteries, they require a continual input of hydrogen-rich fuel. In essence, the fuel cell is a reactor that combines hydrogen and oxygen to produce electricity, heat, and water. It is clean, quiet, and emits no pollution when fed directly with hydrogen.

At the moment, the fuel cell technology is still not cost effective for the commercial building market. But, there seems to be a general feeling that hydrogen-based energy reactors will soon be an optional energy source.

6. *Biogas*

Biogas is produced through a process that converts biomass, such as rapid-rotation crops and selected farm and animal waste, to a gas that can fuel a gas turbine. This conversion process occurs through anaerobic digestion—the conversion of biomass to gas by organisms (like bacterium) in an oxygen free environment.

Biogas has several advantages: it is a relatively high energy production; it lends itself to both heat and power production; it creates almost zero carbon dioxide emissions; it virtually eliminates noxious odors and methane emissions; and it protects ground water and reduces the landfill burden.

7. *Small Scale Hydro*

Harnessing the energy from moving water is one of the oldest energy production systems in the world. In some locations, small scale hydro power is a efficient and clean source of energy and is devoid of environmental penalties associated with large scale hydro projects. It allows small scale, local energy production, with relatively low cost.

8. *Ice storage Cooling Systems*

One of the problems for energy supply companies is that the highest demand for electricity often coincides with the highest cooling demand.

The utilities would prefer to "flatten the curve" (to even out or flatten the measure of average daily energy demand). The less number of peaks (high points of energy demand), the less the utilities have to supplement their power supply with expensive, supplemental fuels.

One way to reduce this peaking problem is to supplement a building's cooling capacity with an ice storage system.

An ice storage system has three components: a tank with liquid storage balls, a heat exchanger, and a compressor for cooling.

The essence of the ice storage system is that the chilling and freezing of the ice balls occurs at night (when the cost of energy is lower due to lower demand). During the day, the cool temperatures, stored in the ice, are transmitted into the building's cooling system.

Source: *Sustainability at the Cutting Edge*, Peter F. Smith, Architectural Press, An Imprint of Elsevier Science. Linacre House, Jordan Hill, Oxford OX2 80P, 2003.

SUMMARY

We have considered the different approaches to heating and cooling buildings, including various types of plants and distribution systems. We discussed the refrigeration cycle and different ways that it may be used. We looked at different distribution systems: electric, hydronic and forced air, and what is involved in sizing forced air systems.

The terms of HVAC equipment are no longer foreign to us, and we have some idea of how energy codes are developed.

This information, along with the information on thermal processes, human comfort, climate, and solar design covered in the previous lessons represent the basic concepts of HVAC on which you may be tested in the examination, and even more important, what we really ought to know to design good buildings.

LESSON 3 QUIZ

1. Which of the following systems can be combined with a four-pipe hydronic system?
   A. Double duct system
   B. Multizone system
   C. Fan coil system
   D. Unitary system

2. All of the following are capable of reversing the otherwise invariant flow of heat from hotter to cooler objects, EXCEPT
   A. the heat pump system.
   B. the variable air volume system.
   C. the refrigeration cycle.
   D. the evaporative chiller.

3. Which of the following systems can provide simultaneous heating and cooling to different rooms within the same zone?
   I.   Single duct constant volume with electric reheat system
   II. Double duct system
   III. Fan coil system
   A. I and II   C. I and III
   B. II and III   D. I, II, and III

4. Which of the following uses the latent heat of evaporation to transfer heat?
   A. Evaporative chiller
   B. Multizone system
   C. Unitary system
   D. Variable air volume system

5. Which of the following systems can be used if each zone must have a separate utility bill?
   I.   Unitary system
   II. Heat pump system
   III. Double duct system
   A. I and II   C. I and III
   B. II and III   D. I, II, and III

6. Which system uses a heat sink and often balances itself without utilizing a boiler or chiller?
   A. Unitary system
   B. Multizone system
   C. Heat pump system
   D. Variable air volume system

7. Which part of the refrigeration cycle is the hottest?
   A. Evaporator   C. Condenser
   B. Compressor   D. Valve

8. Which of the following hydronic systems are parallel?
   I.   One pipe
   II. Two pipe
   III. Three pipe
   IV. Four pipe
   A. I and II   C. III and IV
   B. II, III, and IV   D. I, II, III, and IV

9. Which of the following hydronic systems can provide both heating and cooling simultaneously?

   I.   One pipe

   II. Two pipe

   III. Three pipe

   IV. Four pipe

   A. I and II

   B. II, III, and IV

   C. III and IV

   D. I, II, III, and IV

10. Where does return air go?

    A. To the plant

    B. To the outside of the building

    C. To the room which requires air at that temperature

    D. To the exhaust flue

LIGHTING

INTRODUCTION

We have discussed the thermal processes in buildings and the associated systems. The processes that deal with light are similar in that they are natural processes, yet different in that our eyes are much more sensitive than our thermal senses, and the visible spectrum is a very narrow part of the radiation spectrum.

LIGHT AS THE DEFINER OF ARCHITECTURE

Architects use light in two ways. First, lighting, whether natural or artificial, allows us to see so that we can perform our tasks. In that sense, lighting makes a space usable. But in addition, forms and spaces themselves are perceived in terms of light. How we feel about a building, whether the concept is communicated or the sculptural nature of the building is appreciated, is also determined by light. All of the great architects understood this. Aalvar Aalto, Louis Kahn, and Le Corbusier spoke of light in nearly reverent terms, and made it a key factor in their buildings.

Perception and the Eye

We define light as that part of the electromagnetic radiation spectrum which can be perceived by the human eye. This ranges from blue light, at wavelengths around 450 to 475 nanometers (a nanometer is one millionth of a millimeter), through green and yellow light (at 525 and 575), to red light (at 650). White light is the combination of all of the wavelengths. When we see a wall surface as blue, what really happens is that the white light shines on the wall, and all of the wavelengths except the blue are absorbed by the wall. The blue wavelength bounces back, and is sensed by the eye.

The eye is composed of a focusing device, the lens; a device which controls the amount or brightness of the light admitted to the eye, the iris; and a sensing surface called the retina. The retina is composed of two types of nerve pickups: the cones, which sense colors, and the rods, which sense black and white. The rods work efficiently at very low light levels, such as moonlight. The cones give more information, but require more light. In a very dark room, you lose the sense of color, even if you can still see well enough to move without bumping into things.

The eye is astoundingly adaptive. It can adjust from levels below 1 footcandle to levels over 10,000 footcandles in moments. It is only damaged when the change is too rapid, or when most of the background is dark, but one spot is intensely bright. Such extreme contrasts are known as *glare*.

Perception and the Mind

Incoming information from the eye is analyzed by the mind, which sorts it and interprets it. A good example is our depth perception. Because our eyes are separated, there is a slight difference between what each of them sees. Our brain compares the two images, from which it determines the distance to the object seen. The mind also sorts foreground from background using perspective clues and color clues. Parallel lines approach each other in the distance. Warmer and

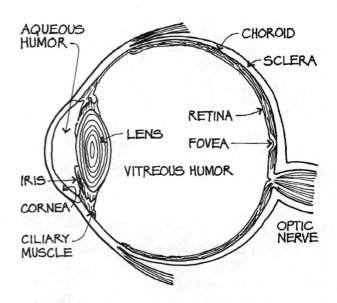

THE EYE

brighter colors tend to be interpreted as nearer, while cooler and darker colors are interpreted as more distant. Shadow information is processed to conclude the shape and form of the objects viewed. Our mind gives us a three-dimensional interpretation of what is really only two-dimensional information arriving at the eye. Sometimes the brain misinterprets this information and may confuse foreground and background, pattern and objects, or it may be fooled, as in optical illusions.

CONCEPTS AND TERMS

Several concepts and terms which are commonly used must be defined.

Transmission, Reflection, Refraction and Absorption

All light which strikes a surface is either transmitted, reflected, or absorbed. *Transmitted* light passes through the material. If the *image* is transmitted, the material is called *transparent*. However, material which is transparent may

POLISHED SURFACE
SPECULAR

ROUGH SURFACE
SPREAD

MATTE SURFACE
DIFFUSE

DIFFUSE AND SPECULAR REFLECTION

change the image, such as the lens on a pair of glasses. This is called *refraction*, and occurs to some extent with nearly all transparent materials. When one looks at a fishing line in clear water, the line seems to bend sharply just below the surface. It is not the line that is bent, but the path of the light rays from the line, through the water and air interface.

If no image at all is transmitted, but there is still light passing through the material, it is called *translucent*. Frosted glass is an example. If the light is bounced off the material, it is called *reflective*. If the reflected image is maintained, as with a mirror, the surface is called *specular*. If the image is not maintained, as with a matte white finish, the surface is called *diffusing*. If absolutely no light passes through the material, it is called *opaque*. All of the light is either reflected or absorbed, or both.

Direct and Diffuse Light

Light is usually available to us in two forms. *Ambient* or *diffuse* light is the kind of light experienced on an overcast day. There are no sharp shadows, because the light is coming from all directions. In a building, this is analogous to a luminous ceiling, or a white ceiling lit by coves around the sides. This lights the entire room or area, and is therefore referred to as *area* lighting.

Direct light is the kind of light which comes directly from the sun on a sunny day. There are very sharp shadows, and the light is very strong. There are also very distinct reflections from shiny objects. Direct light is analogous to the light from a projector, or to a lesser extent, from a drafting lamp. It is most often useful when aimed at a task requiring special attention, and when so used is called *task* lighting.

Flat surfaces, such as murals, paintings, and papers or books, are best viewed in diffuse light, which prevents *veiling reflections* or *reflected glare*. Strongly modeled objects, such as sculptures, are more amenable to dramatic lighting, such as direct lighting, which casts sharp shadows, allowing us to understand the form.

Kelvin and Color Rendition Index

"Perfect" white light consists of a complete spectrum of wavelengths, with an even distribution. However, white light which has been transmitted through a translucent surface or reflected off of a surface is often shifted in color, or missing some part of the spectrum. Similarly, "artificial" light created by bulbs, tubes, or lamps sometimes has parts of the spectrum missing, or may have the distribution shifted one way or another. The measure of how well light actually shows true color is called the *color rendition index (CRI)*. This term is most often

<u>Actual Size</u>

Top Lighting

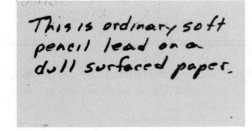

Side Lighting

<u>Five Times Size</u>

Top Lighting

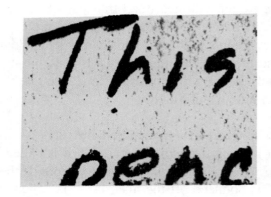

Side Lighting

At this size, you can see the actual reflections which break up the image.

<u>VEILING REFLECTIONS</u>

Used by permission of Pilkington North America, Inc.

employed with artificial lighting. The best possible rating is 100, in which there are no colors missing. Another way of rating white light is called the *color temperature*, which comes from the theoretical relationship between the temperature of an object and the color of the light emitted, e.g., red hot to white hot to daylight (the surface of the sun is at approximately 6,000° Kelvin). The filaments or phosphors in the light source are not necessarily at the

temperature indicated, but the color of the light is described nonetheless. We will discuss this in more detail when we cover artificial lighting.

BASIC PHYSICS

In everyday English, we use many lighting terms interchangeably and often imprecisely. In dealing with lighting in buildings, however, our terminology must be precise, and therefore candidates should become familiar with the definitions which follow.

Power or Intensity

The amount of light put out by a source is called the *intensity (I)*. The unit of measure for the intensity of a source is the amount of light coming from a single candle, called one *candlepower (cp)*.

Flux

If we drew a one foot square in midair, at a distance of one foot from a one candlepower source, the amount of light flowing through that square would be termed one *lumen (I)*. This flow through a theoretical surface is called the *flux (F)*.

Illumination

If we were to place a candle one foot from a blackboard there would be one lumen arriving on one square foot of blackboard surface. This is the *illumination (E,* since we already used I), and its value would be called one *footcandle* (fc). $E = F/A$.

Luminance

Last of all, the amount of light leaving the surface of the blackboard would depend on its reflectivity and would give us a measure of how bright it looked, or its *luminance*. For example, a perfectly reflective surface exposed to an illumination of one footcandle would have a luminance of one *footLambert (fL)*. Brightness, or luminance, may also refer to the amount of light passing *through* a translucent surface. For example, a white surface with a reflectance of 80% and a white material with a transmittance of 80% will both have the same brightness

if exposed to the same illumination. The translucent surface will simply have that brightness on the far side, rather than the near side.

Inverse Square Law

Several basic rules may be used in lighting calculations. If the source of light may be approximated as a point (a candle, a light bulb, even a single tube or fluorescent fixture), the flux, and resultant illumination, is inversely proportional to the square of the distance from the surface.

For example, if a lamp has an intensity of 1,600 cp, a perpendicular surface 10 feet away has an illumination of:

$$E = I/d^2 = 1,600 \text{ fc}/(10 \text{ ft})^2$$
$$= 16 \text{ fc}$$

If we double the distance to the surface, to 20 feet, the illumination will be:

$$E = I/d^2 = 1,600 \text{ fc}/(20 \text{ ft})^2$$
$$= 4 \text{ fc}$$

Thus, doubling the distance cuts the illumination to one quarter. This can also be expressed by the relative formula:

$$E_2 = E_1(d_1/d_2)^2$$

Thus the same example becomes:

$$E_2 = E_1 (d_1/d_2)^2 = 16 \text{ fc } (10ft/20ft)^2$$
$$= 4 \text{ fc}$$

We can examine the brightness of surfaces in the example above. If we place the lamp 10 feet from a white wall with a surface reflectance of .75, the brightness or luminance of the wall is

$$L = 16 \text{ fc} \times 0.75 = 12 \text{ fL}$$

If we place a piece of frosted glass with a transmittance of .75 at 10 feet from the lamp, the brightness or luminance is:

$$L = 16 \text{ fc} \times 0.75 = 12 \text{ fL}$$

LIGHTING SYSTEMS

Different artificial sources produce different kinds of light, and vary significantly in their efficiency (or efficacy, which is the calculated lumen output per watt input). There are three general categories: incandescent, fluorescent, and high intensity discharge.

Incandescent

An incandescent lamp or bulb contains a *filament,* usually a tungsten alloy, which is heated by passing an electric current through it. It glows, giving off light and a significant amount of heat. The gas in the lamp is inert, such as nitrogen or argon, and does not interact with the filament or corrode it. Incandescent light is typically "warmer" than sunlight or daylight, rich in yellows and reds, and weak in greens and blues. Because much of the energy is wasted in the production of heat, incandescent lighting is the least efficient type of artificial light. It is also characterized by a short lifetime for individual bulbs. An output of *15 to 18 lumens per watt* is common, and a lifetime of around 2,000 hours is typical. The lifetime and output of bulbs are inversely related. Thus, burning a lamp at a lower voltage results in less light and a "warmer" color, but increases the lifespan of the lamp.

Incandescent lamps come in various shapes, with different characteristics. The most common shape is known as the A shape, which is the kind found in table lamps. Incandescent bulbs are sized in terms of wattage, and in multiples of 1/8 inch in diameter. For example, a 100 watt A-19 bulb uses 100 watts and is 19 × 1/8 inch, or 2.375 inches in diameter. R and PAR lamps have an internal reflector so that all of the light comes out of the front of the lamp. These are both more expensive and more effective than an A lamp at lighting a specific object. There are lamps with and without diffusing surfaces. There are lamps which run at lower voltages, such as 12 or 24 volts, which allows a smaller filament and results in a better focus of the beam.

Tungsten-halogen lamps are incandescent lamps which house the filament within an inner quartz envelope which can tolerate higher operating temperatures. This also contains a special halogen gas which prevents the evaporated metal from the filament from depositing on the inner surface of the quartz. Thus, the filament can be run at much higher temperatures, which produces more light and a slightly better color. In addition, the redeposition of the metal back onto the filament extends the life of the lamp slightly.

Fluorescent

Fluorescent lighting is a much more efficient system based on passing a current through gases inside a glass tube. This releases energy in the form of free electrons and gas ions. The glass tube can be lined with phosphors which in turn are excited by the ions, and which then glow in characteristic colors. If the phosphors are combined correctly, good color renditions can be achieved.

Unfortunately, it is impossible to get the current to arc through the gas at 110 volts, and therefore a transformer is necessary. In addition, once the arc has been formed, the resistance of the tube changes, and the circuit must be

INCANDESCENT LAMP SHAPES

adjusted to avoid excessive current. A fluorescent fixture actually consists of the lamp and an associated *ballast* which controls the voltage and the current to the lamp. Ballasts are sometimes noisy, and are assigned a Sound Rating from A to E, A being the quietest.

Four foot lamp lengths utilizing 40 watts or less are most common, although there are now small U-shaped lamps five inches in length, as well as large high-output lamps up to eight feet in length. There are a myriad of color combinations available. *Cool white* produces the most lumens per watt, but its color is unflattering to most people's skin tones. *Warm white* is somewhat better. *Cool white deluxe, warm white deluxe, royal white*, and the *SP* series lamps provide better CRI at slightly greater costs, because of the rare earth phosphors required. The lifetime of a fluorescent tube is determined not only by the number of hours it is lit, but also by the number of times it is switched on and off, which tends to wear it out. Based on a three-hour burning perod every time the lamp is turned on, the average lifetime is about 10,000 hours. Efficiencies usually range from *60 to 80 lumens per watt.*

High Intensity Discharge

High intensity discharge (HID) lamps consist of a lamp within a lamp, which is run at a very high voltage. There are four general types, only three of which are used in architectural lighting. The first HID lamps were *mercury vapor* lamps. They produce a very bright, clear, bluish light which unfortunately makes humans look unhealthy. This can be improved by adding phosphors to the inner surface of the outer envelope. Such lamps are called *mercury vapor deluxe.* Lifetimes are in the 24,000 hour range, and output is up to *50 lumens per watt.*

The next improvement in HID lamps was the inclusion of a *metal halide* gas, typically iodine, in the inner envelope, which shifted the color and improved the efficacy to approximately *80 lumens per watt.* However, the expected lifetime for the lamp dropped to about 10,000 hours.

The most efficient of the architectural HID lamps is the *high pressure sodium (HPS)* lamp. It develops up to *110 lumens per watt,* and has a 24,000 hour life expectancy. Unfortunately, the color rendition suffers somewhat, although new coated "deluxe" models seem to be a vast improvement at a very slight drop in efficacy.

Low pressure sodium lamps have the highest ratings in both lifetime and efficacy, but produce a monochromatic yellow light suitable solely for security lighting, since there is no color rendition at all. Everything is seen as if it were black and white, only it is black and yellow instead. Red looks black, blue looks black, and white and yellow look identically yellow.

TUBE CONTAINS RARE GAS AND MERCURY VAPOR

ELECTRODE COATED WITH ELECTRON·EMISSIVE MATERIAL

BASE CEMENT

BASE PINS

EXHAUST TUBE

STEM PRESS

MERCURY

<u>RAPID - STARTING</u>

<u>INSTANT- STARTING</u>

FLUORESCENT TUBES

The least efficient lamp type is normal incandescent. Tungsten-halogen is a slight improvement, followed roughly by mercury vapor, then fluorescent and metal halide in a tie, and finally the most efficient, high pressure sodium. Low pressure sodium is even more efficient, but can only be used in security lighting or similar applications.

MAIN ELECTRODE

QUARTZ ARC TUBE

STARTING ELECTRODE

OUTER GLASS ENVELOPE

<u>HID LAMP</u>

ARTIFICIAL LIGHTING CALCULATIONS

There are two common methods for calculating light levels. One works best for a single fixture, or small numbers of fixtures, and is called the *point grid* method. It takes very careful account of orientation and distance, but ignores surrounding reflection. The *zonal cavity* or *room cavity* or *lumen* method is based on a uniform distribution of a large number of fixtures, and takes into account the reflectivity of the ceiling and walls, and the comparative volumes of the top, middle, and bottom of the room. Both will be covered briefly, since detailed calculations are not likely on the examination. The Illuminating Engineering Society (IES) publishes an excellent reference manual for use when detailed calculations are necessary.

Point Grid Method

The point grid method is based on the formula:

$$E = I \cos \theta / d^2$$

where

    E = illumination at the receiving surface

    I = intensity at the source when viewed from the direction of the receiving surface

    θ = the angle between a perpendicular (normal vector) to the receiving surface and a line from the source to the surface

    d = the distance from the source to the surface

ILLUMINATION (E) (ON THE HORIZONTAL PLANE)

$$= \frac{\text{CANDLEPOWER} \times \cos \theta}{D^2}$$

POINT GRID METHOD

CANDLEPOWER DISTRIBUTION CURVES

The intensity in a given direction is taken from polar plots of the fixture intensity sometimes known as *candlepower distribution curves.* These graphs show how much light is given off at any given angle from a vertical reference line. If all of the light goes up, the fixture is called an *indirect* fixture, because the light will reach the work plane only after bouncing off the ceiling. This is particularly suitable for uniform light in rooms with computer terminals. If all of the light goes down, it is called a *direct* fixture.

If the light spreads widely to either side, it gives a uniform illumination on the floor, but may cause glare by shining into the user's eyes even when he or she is not looking up. If the beam spread is narrow, the fixtures must be closely spaced to avoid spotty illumination on the floor and work plane.

Abney's Law simply states that the light arriving at a surface is the sum of the light arriving from all of the sources, and can be expressed by repeating the point grid formula for each source:

$$E = I_1 \cos\theta_1/d_1^2 + I_2\cos\theta_2/d_2^2 + . \pm I_n\cos\theta_n/d_n^2$$

This means that we can add the light from all of the fixtures in a room, one at a time, and get the correct result for all of the direct light, but the reflected light is not considered.

Zonal Cavity Method

The method most commonly used for office, commercial and factory spaces is the zonal cavity method. It is based on a coefficient of utilization (CU) for each fixture type which takes into account the direction that the fixture throws its light, the reflectances of the ceiling cavity, the middle level of the walls, and the zone between the work surface and the floor. The CU theoretically varies between 0 and 1.0, with most values in the .5 to .8 range. The formula also considers maintenance factors and dirt, and is expressed by the equation:

$$E = (N \times n \times LL \times LLD \times DDF \times CU)/A$$

where
   E = the illumination in footcandles
   N = number of fixtures
   n = the number of lamps per fixture
   LL = number of lumens produced
          per lamp
   LLD = the lamp lumen depreciation factor
          (which accounts for the effects of
          aging on lamp output)
   DDF = the dirt depreciation factor for the
          fixture based on scheduled mainte-
          nance and how dirty the surroundings
          are
   CU = the coefficient of utilization, calcu-
          lated as discussed in the paragraph
          above

A = the area of the working plane (or floor) which will be illuminated by the fixtures

The equation can be manipulated so that if the designer knows the desired illumination level in footcandles (E), he/she can solve for the necessary number of fixtures (N). The result is

$$N = E \times A/n \times LL \times LLD \times DDF \times CU$$

RECOMMENDED ILLUMINATION

The amount of light required to see well varies greatly with the age of the observer and the task at hand, and there is significant disagreement about the ideal illumination levels. How diffuse the light is also affects these levels. The optimum lighting is called the *equivalent spherical illumination (ESI)*. It is based on a theoretical sphere surrounding the object being illuminated with the light cast evenly from all parts of the sphere, eliminating any shadows and any reflected bright spots. 100 footcandles of light from a single lamp may yield a level of legibility equivalent to only 50 footcandles ESI. The concept is important, but the calculations are beyond the scope of this course.

DAYLIGHT CALCULATIONS

Perhaps the most significant rediscovery of the past ten years is the use of natural daylighting to save energy and improve the quality of light.

Whenever we can turn off a fixture, we reduce the electrical load and reduce the heat generated within the space as well. In hot climates this

HIGH SUMMER SUN WARMS THE ROOF AND EAST & WEST WALLS.

LOW WINTER SUN WARMS SOUTH WALL

OVERHANG ALLOWS WINTER SUN TO ENTER BUT BLOCKS OUT SUMMER SUN.

SOUTH

THE SUN'S LOCATION VARIES WITH THE SEASONS

results in a double benefit. Even in cold climates, there is often an economic benefit. In addition, daylight is a diffuse light source, and the color rendition is perfect.

However, certain precautions should be taken. Direct sunlight should be shielded, diffused by diffusing glass, or bounced off of diffusing surfaces. Horizontal skylights should be introduced with care, because of the potential for great heat gain in the summer and little heat gain in the winter. (Remember that horizontal surfaces get *more* sunlight in summer and *less* sunlight in winter.)

Artificial lighting should be provided even in daylit areas, for nighttime illumination. Lighting fixtures may be connected to switches in such a manner that fixtures can be turned off in areas where daylight is sufficient, while other fixtures may be left on. Even better, dimmer controls may be connected to a photosensitive device which dims the lights successively as the daylight level increases. Both stepped and continuously dimmable switches and ballasts are now available for fluorescent fixtures.

Daylighting Strategies

There are several architectural elements which can be used in daylighting strategies.

A *light shelf* is an overhang with glass above it, which reflects light into the room and up onto the ceiling. Such light shelves are usually placed above head height, to avoid reflecting glare into the eyes of the occupants. A *glass transom* is a transparent area over a door, or even over shelves, bookcases, etc., which lets light pass from one room into another, while providing some security and acoustical privacy. A *saw-tooth* is a roof with a series of vertical or nearly vertical glass surfaces, usually facing north. This lets in the even, diffuse light, but limits the direct sunlight to early summer mornings and late summer evenings. For more information on daylighting techniques, check some of the references in the bibliography.

There are two methods for calculating the amount of daylight in a space based on orientation, windows, and internal and external reflection.

Lighting and Sustainable Design

The illumination of the interior of a sustainably designed building requires a holistic approach that balances the use of artificial and natural lighting sources.

1. *Daylighting*

   Properly filtered and controlled solar radiation may provide a valuable source of illumination to a building interior. This process is called "daylighting" (simply having properly designed fenestration that allows natural sunlight to replace or dramatically reduce the need for artificial lighting).

   Because unwanted sunlight (particularly in summer months) can also add to the internal heat load of a building, the architect must be careful to allow only beneficial sunlight and reduce unwanted solar heat gain. There are several techniques for controlling daylighting:

   - Overhangs, fins, and other architectural shading devices.

   - Sawtooth (not bubble) skylight design, which allows the glass to face north for illumination—not south for solar heat gain.

   - Interior window shading devices, which allow solar gain during cool months, and the blocking of solar radiation during the warmer seasons.

   - Light shelves, which permit the daylight to reflect off the ceiling and penetrate farther into the interior without affecting views outside.

2. *Higher Efficiency Light Fixtures*

   In addition to a daylighting strategy, light fixtures that are more efficiently designed reduce energy cost and increase comfort, such as:

   - Fixtures that use fluorescent or HID lamps, which provide more illumination per watt than incandescent lighting.

   - Fixtures that are designed to diffuse or bounce the illumination off the ceilings or internal reflectors, which are more efficient, cause less glare (especially in an environment with computer monitors); and save operating costs.

   - Fixtures that have higher efficiency (T-8) fluorescent bulbs, which produce more lumens per watt and thereby diminish the heat generated by lighting.

   - Fixtures that offer dimming or multiple switching capability, which permit the architect a more energy efficient lighting design.

   Dimming or multiple switching fixtures allow the architect to design lighting patterns that blend nicely with daylighting opportunities.

   - For example, an office with perimeter fenestration allows daylighting supplemented with overhead lighting that can be dimmed or reduced.

   - The interior spaces, which are too far from the perimeter for daylighting, may be controlled with switches or dimmers that allow relatively higher levels of illumination.

   - The result is an even illumination pattern, which saves on artificial lighting costs, by relying on daylighting at the perimeter.

   - Fixtures that use higher efficiency lamps such as fluorescent, high intensity discharge (HID) sulfur lighting (exterior only).

   - Fluorescent fixtures that use high efficiency electronic ballasts.

Additionally, the architect may avoid less efficient incandescent lighting where possible, install task lighting to supplement diffused ambient lighting, and install LED (light emitting diode) lighting for exit signs. LED lighting lasts longer than incandescent and is far less expensive to operate.

3. *Lighting Sensors and Monitors*

Where possible, lighting cost can be diminished by installing light monitors that sense occupancy conditions. As long as the room contains people, the lights will remain on. If people leave, the sensor will wait for a few minutes, then shut off all the lighting in the room.

Lighting sensors can be designed to operate with a preference for motion, heat (from people), or desired time of occupancy.

4. *Lighting Models*

Computer lighting models are one option that allows the architect to simulate the levels of sunlight that penetrate into a building design, depending on the building location, varying times of year, fenestration orientation, and design.

By incrementally altering fenestration (skylights, windows, or light transport systems) and the artificial lighting system, the architect may optimize the daylighting and artificial lighting systems for the building.

Lumen Method

The method developed in the United States is called the lumen or flux method. It is approved by the IES and was long promoted by Libbey-Owens-Ford as a public service. It is well suited to both clear skies and partly cloudy skies. The amount of daylight is calculated in only three locations in the room: 5 feet from the window, the middle of the room, and 5 feet from the back of the room. It can be used to calculate the amount of daylight from one window wall, or two window walls on opposite sides, but not from a corner window (windows on two adjacent walls).

Daylight Factor Method

The other method is called the daylight factor method, which was developed in Europe and assumes overcast or diffuse sky conditions. This method is sanctioned by the Commission Internationale d'Eclairage (CIE) and is often used in computer programs. The major benefit is that it can be used to calculate the amount of daylight at any location in the room, including the effect of corner windows.

The result of a daylight factor calculation is a number which expresses the amount of light at a particular interior location as a percentage of the light which is available on an exterior horizontal surface. For example, a daylight factor of 3 at a corner of the room means that 3 percent of the light available outside would arrive on the workplane in that corner. Since outside light levels can be quite high, 3 percent might be sufficient. If there were 2,000 fc on the ground outside, then a DF of 3 would result in an illumination level of $2,000 \times 0.03 = 60$ fc.

EMERGENCY AND EXIT LIGHTING

Most building codes require special emergency lighting at exits and for certain critical functions. Emergency lighting may be provided by a separate generator in the building, or by battery packs and small lights which would provide enough illumination to evacuate the building. Most codes do not allow lead-acid battery packs because of the fumes emitted, although there are exceptions with small batteries or proper ventilation. *Nickel-cadmium* batteries are slightly more expensive, are rechargeable, and emit no noxious fumes. For fluorescent emergency lighting, some kind of *transformer* and *inverter* are necessary, since fluorescents do not run on 12 volt DC current.

Exit signs themselves must be illuminated by at least two sources: the general illumination and an internal or special illumination. Even with the lights out and the room full of smoke, the exit signs should be visible to escaping occupants.

SUMMARY

We have studied light, the definer of space and form, starting with some of the processes involved in perception, and then the terms and physics of lighting. Different types of artificial lighting have different characteristic CRI's and efficacies. There are different ways of calculating illumination, both for single point sources and for uniformly lit rooms, as well as for natural lighting in clear and overcast skies. All of these factors together yield an understanding of lighting within buildings, as necessary to the examinee and to the architect. For further study, the Illuminating Engineering Society's *1984 Reference Volume* is recommended.

LESSON 4 QUIZ

1. Which color represents the highest Kelvin temperature and the shortest wavelength?
   A. Red
   B. Yellow
   C. Green
   D. Blue

2. What is the name of the sensing surface of the eye?
   A. The lens
   B. The iris
   C. The retina
   D. The cilia

3. Which of the following statements about glare are true?
   I.   Glare is always a reflection.
   II. Glare is always an extreme contrast in brightnesses.
   III. Glare is always a very high illumination.
   A. I only
   B. II only
   C. III only
   D. I, II and III

4. A material which transmits light without transmitting the images on the far side is called
   A. transparent.
   B. translucent.
   C. opaque.
   D. reflective.

5. Which of the following statements are correct about the term diffuse?
   I.   Ambient light is usually diffuse.
   II. A mirror is a diffuse reflecting surface.
   III. Printing or writing is easiest to read when viewed in diffuse light.
   A. I and II
   B. II and III
   C. I and III
   D. I, II, and III

6. What is the unit which measures the amount of light striking a surface (the illumination)?
   A. Candlepower
   B. Lumen
   C. Footcandle
   D. FootLambert

7. If a lamp of 2400 cp intensity results in an illumination of 24 fc at 10 feet, what is the illumination at 20 feet?
   A. 6 fc
   B. 8 fc
   C. 10 fc
   D. 12 fc

8. The major advantage to low voltage incandescent lamps is that the lamps
   A. are much brighter for a given wattage.
   B. last much longer.
   C. focus and aim their light more accurately.
   D. can run at higher temperatures.

9. What is the diameter of an R-40 bulb?
   A. 2.5"
   B. 3.2"
   C. 4"
   D. 5"

10. Which of the following are true about fluorescent lighting?
    I.   Fluorescent lighting is more efficient than incandescent lighting.
    II. Fluorescent lighting can have different CRI ratings, depending on phosphors.
    III. Fluorescent lighting does not run on 12 volt DC current.
    A. I and II
    B. II and III
    C. I and III
    D. I, II, and III

ELECTRICAL SYSTEMS

INTRODUCTION

Although the use of electrical power has little effect on the form or shape of buildings, it is of tremendous importance when it comes to the tasks performed within buildings. In fact, without the electricity which powers its typewriters, air conditioning, lighting, and computers, the modern office building would be unable to function. The architect should therefore have an understanding of the basic principles of electrical systems, and the Architect Registration Examination tests this understanding.

BASIC PHYSICS

Because we cannot see electricity, perhaps the best way to understand it is by looking at the flow of water as an analogy.

The three basic factors in electricity are *potential, current*, and *resistance*. The chart on the following page shows the analogous factors in water flow.

	Water	Electricity
Potential	Height, pressure or pressure difference (feet or psi)	Voltage V (volts)
Current	Flow (gallons per minute)	Current I (amperes)
Resistance	Resistance to flow (inches per 100 ft.)	Resistance R (ohms or Ω)

The formula which relates these factors to each other is called *Ohm's Law*:

$$I = V/R$$

where

    I = current, measured in amps
    V = voltage, measured in volts
    R = resistance, measured in ohms

The formula makes sense, for both the analogous water flow and electricity. The greater the voltage, the greater the current (the greater the water pressure, the greater the water flow). Conversely, the greater the resistance, the smaller the current.

As with water, there may be several resistances in the flow path, or there may be parallel paths with different resistances and flow rates in each path. In electricity, these two conditions are called respectively *series resistances* and *parallel resistances*. The effective resistance of each may be calculated.

With series resistances, the effective total resistance is the sum of all of the resistances:

$$R_{tot} = R_1 + R_2 + R_3 + \cdots + R_n$$

With parallel resistances, the effective total resistance is expressed by the formula:

$$1/R_{tot} = 1/R_1 + 1/R_2 + 1/R_3 + \cdots + 1/R_n$$

$$R = R1 + R2 + R3 + R4$$
$$R = 5 + 10 + 15 + 20 = 50 \text{ ohms}$$

SERIES RESISTANCES

$$R = \frac{1}{1/R1 + 1/R2 + 1/R3 + 1/R4} =$$
$$\frac{1}{1/5 + 1/10 + 1/15 + 1/20} = 2.4 \text{ ohms}$$

PARALLEL RESISTANCES

TRANSMISSION AND USAGE

There are several schemes for using and transmitting electricity, the most common of which are described below.

Direct Current

Direct current (DC) means current which flows only in one direction, with constant voltage. This is typical for low voltage applications, as with batteries, for example. Low voltages are less dangerous, in general, than higher voltages, because they result in less current through a given resistance. The general equation for power in a DC circuit is:

$$P = V \times I$$

where

P = power in watts

V = voltage

I = the current in amps

For example, for a 12 volt battery connected to a 4 ohm resistor, the current is

$$I = V/R = 12v/4\Omega = 3 \text{ amps}$$

$$P = V \times I = 12v \times 3 \text{ amps} = 36 \text{ watts}$$

Alternating Current

Alternating current (AC) is based on the concept that electricity has nearly no inertia, and therefore the direction of the flow can be reversed very rapidly by reversing the voltage. If we plot the voltage in such a circuit, it results in a sine wave.

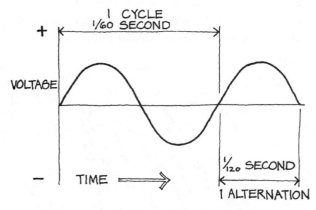

ALTERNATING CURRENT SINE WAVE

The current flow may lag behind the voltage reversal, and therefore the amount of power is not as simple to calculate as with a DC circuit. The *power factor* is the cosine of the angle between the voltage wave and the resultant current wave. It ranges from 0.0 to 1.0, although it is usually expressed as a percentage, i. e., from 0 to 100%. In a single-phase circuit, the power formula is:

$$P = V \times I \times PF$$

where

P = power in watts

V = voltage in volts

I = current in amps

PF = power factor in decimal form

There is also a three-phase version of alternating current, in which there are three different circuits, each 120° out of phase with the others, and one neutral or ground circuit. The formula for power in a three-phase circuit is:

$$P = V \times I \times PF \times \sqrt{3}$$

where

P = power in watts

V = voltage in volts

I = current in amps

PF = power factor in decimal form

For example, if a three-phase motor draws a current of 7 amps at 240 volts and the power factor is 0.8, the power

$$\begin{aligned} P &= V \times I \times PF \times \sqrt{3} \\ &= 240 \text{ v} \times 7 \text{ amps} \times 0.8 \times 1.73 \\ &= 2,325 \text{ watts} \end{aligned}$$

When we deal with a large amount of power, we commonly use the unit *kilowatts* (1,000 watts) or even megawatts (1,000,000 watts). Thus 2,325 watts is the same as 2.325 kilowatts.

ELECTRICAL EQUIPMENT

A motor is a machine which converts electrical energy into mechanical energy. The converse is a generator, a machine which converts mechanical energy into electrical energy.

Rotating a wire loop between two magnetic poles will generate a current. This is the basic principle behind *generating* electricity.

Conversely, running a current through a wire wrapped around an iron core creates a magnetic field. Since magnets may attract or repel each other, the magnetic field can create motion. This is the basic principle behind electric *motors* and electric *solenoids*.

A solenoid is a wire wound spirally around an iron core to produce a magnetic field and which is used as an electromagnetic switch.

Generation of Power

A single-phase *alternator* is the most basic form of power generation. The resultant power is AC current, in which the time interval from peak to peak of the voltage sine wave is based on the number of revolutions per minute (rpm) of the shaft on which the wire loop is mounted. This is usually 60 rpm, which results in a peak to peak time (one cycle) of 1/60 of a second, or 60 cycles per second or 60 *Hertz*, the typical AC power frequency in the United States. In Europe, 50 Hertz is the common frequency for power. In the United States, the common household voltage is 110 volts, while in Europe it is 220 volts. This is the magnitude from the bottom to the peak of the voltage sine wave.

Three-phase power is generated by putting three loops on the shaft and keeping them as separate circuits. If the loops are evenly spaced around the circumference of the shaft, the sine wave current generated is shifted by 1/3 of a cycle (or 120 degrees) between each circuit.

The resulting currents can be represented by three separate sine waves. Note that if only one of the circuits is connected, normal single-phase current is delivered.

Transformers

Devices which change the voltage of an AC circuit to a higher or lower value are called

SINGLE PHASE ALTERNATOR

transformers. Basically, a transformer consists of an iron core on which two separate coils of wire are wound. The coil (also called a winding)

THREE PHASE ALTERNATOR

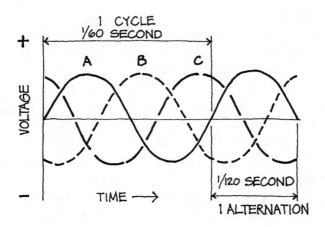

THREE PHASE CURRENT

with the greater number of turns will have a higher voltage, and the one with fewer turns will have a lower voltage. We can run current through the winding with the greater number of turns and produce a lower voltage in the other winding, or vice versa. While a transformer changes the voltage in a circuit, it has practically no effect on the total power in the circuit. Transformers are used to step up voltage in order to transmit power over long distances without excessive losses, and subsequently step down voltage to more usable household levels.

Transformers are called *step up transformers* when they increase the voltage and *step down transformers* when they decrease the voltage, which is usually the case in buildings.

TRANSFORMER

Transformers waste surprisingly little energy, but what is wasted turns into heat, which must be dissipated by the transformer. The thermal rating of the transformer is the product of the Voltage and the Amperage or VA. This is not exactly the same as power (V × I × PF) because it represents how much heat the transformer can handle without melting or exploding, not power being delivered. Again, in the scale of buildings, one thousand VA is a handy unit, and is abbreviated KVA.

In small transformers, the wires are simply insulated by rubber or vinyl or other usual insulating materials. In larger transformers, however, the wires are insulated by a fluid which is resistant to the flow of electricity and also capable of withstanding the high internal temperatures, rapidly conducting the heat away from the windings.

Transformers should be properly ventilated, since their thermal rating is based on the assumption that the surrounding air will be able to remove heat at a certain rate from the cooling fins on the transformer. If a large transformer overheats, it may explode. The fluid that insulates the electricity and cools the transformer is often toxic, and the explosion may vaporize it or spray it over adjacent surroundings or people.

For these reasons, transformers must either be placed outside the building, or within the building inside a fireproof vault. In addition, transformers make a certain amount of noise (hum) and it is desirable to isolate that as well.

Transformer Connections

The *primary* winding is the winding which is used for input in a transformer, while the *secondary* winding is used for output. The secondary may be divided into segments, so that the output voltage depends on which segments

TWO WIRE SECONDARY

are used. Single-phase transformers may have two- or three-wire secondaries.

A *two-wire secondary* has one wire grounded, which then becomes the neutral.

A *three-wire secondary* consists of two segments. One lead is at one end of the secondary. A second lead is connected to the midpoint of the secondary and grounded, and the third lead is connected to the other end of the secondary. If 240 volt output is desired, the first and third lead are used, and the midpoint lead is ignored. For 120 volts, the first and midpoint leads or the midpoint and third leads are used.

THREE WIRE SECONDARY

Similarly, three-phase transformers may have multiple leads on the secondary winding, and even different configurations to both the primary and the secondary.

The two basic types of connections are called wye (shaped like the letter Y) or delta (shaped like the Greek capital letter delta Δ, or a triangle). The wye is sometimes referred to as a "star" because the neutral contact (at the crotch of the Y) forms the center of a three pointed star.

The illustration on the following page shows five basic methods for connecting three-phase transformers (the iron core symbols have been omitted from the illustration).

When a neutral connection is desired, it may be taken from the center point of a wye, or from the midpoint of one of the secondary windings of a delta. The illustration on page 96 shows alternative representation of delta-delta and delta-wye transformers with neutral connections added. The parallel lines represent the iron cores, and the neutral wire is shown grounded for safety, which is typical.

Primaries are usually connected delta, and seldom have a ground connection. Since the wye connection is symmetrical, the voltage from each of the three-phase wires (A, B, and C) to the neutral is the same, and is equal to the voltage from line-to-line divided by $\sqrt{3}$ (or 1.73). Typical system voltages are 120/208 and 277/480, where the lesser voltage is the line-to-neutral voltage and the greater voltage is the line-to-line voltage. Note that in both cases the ratio of the voltages is 1.73.

Where a neutral is applied to a delta, the voltage from two of the phase wires (B and C) to the neutral is one half of the phase-to-phase voltage. The voltage from the third phase (A) to the neutral is 0.866 times the phase-to-phase voltage, but no loads are ever connected between these two points.

Residences are limited to 120/240 volt single-phase systems, which come from a three-wire secondary, and larger loads, such as the electric range, air conditioner, refrigerator, and other semi-permanent connections, use the line-to-line voltage of 240 volts. The receptacles and light switches use the line-to-neutral voltage of 120 volts.

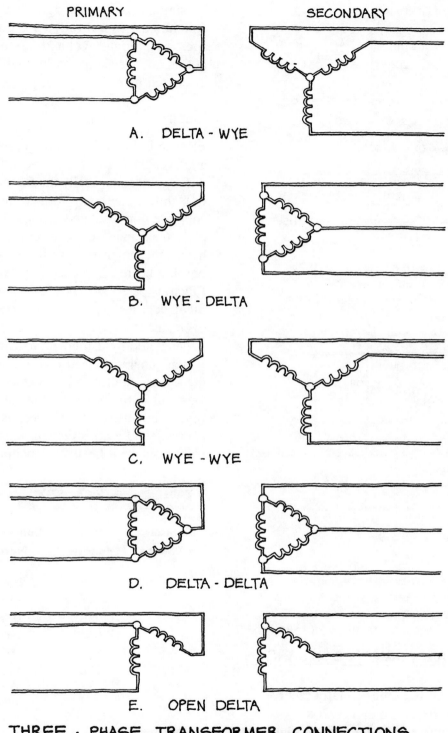

PRIMARY   SECONDARY

A.   DELTA - WYE

B.   WYE - DELTA

C.   WYE - WYE

D.   DELTA - DELTA

E.   OPEN DELTA

THREE · PHASE TRANSFORMER CONNECTIONS

DELTA - WYE SYSTEM
120/208V 3 PHASE 4 WIRES
277/480V HAVE SAME CONNECTIONS

DELTA - DELTA SYSTEM
120V/240V 3 PHASE 4 WIRES
WITHOUT THE NEUTRAL, THIS CAN BE
240V 3 PH 3W OR 480V 3 PH 3W

3·PHASE TRANSFORMER SCHEMATICS

Small commercial buildings usually receive 120/240 volt single-phase or 120/208 three-phase service, but larger commercial and industrial applications require the higher voltage three-phase services, such as 277/480 volts or 2,400/4,160 volts.

To summarize, the voltages in Table 5.1 are in common use as utilization and distribution voltages. Even higher voltages are used for distribution and transmission, but will not be discussed here.

Electric Heaters

Electric heaters, the heating coils in furnaces, and even the coils in hair dryers which warm the airflow are all basically the same thing: a length of stainless steel wire formed into a coil and supported on insulated prongs. The wire is simply a resistance to the current, which generates heat. In a motor, this would be wasted energy, but in a heater, it is exactly what we desire. Electric heaters are 100% efficient, in that every bit of electrical energy is converted to heat. However, since much of the electricity in this country is generated by using heat to generate steam, which generates electricity at about a 30% efficiency, using electricity to heat a space is very wasteful in the broad ecological sense. On the other hand, using electric heat as a *radiant heat* source (see Lessons One through Three) is efficient because it heats only people, and not air.

TABLE 5.1—TYPICAL SECONDARY VOLTAGES			
Number of Phases	Voltage	Connection Name	Number of Wires
1	120	2 wire	2
	120/240	3 wire	3
3	240	Delta	3
	120/240	Delta	4
	480	Delta	3
3	208	Wye	3
	120/208	Wye	4
	480	Wye	3
	277/480	Wye	4
3	2,400/4,160	Wye	4
	4,160	Delta	3
	4,160	Wye	3

Electric Lighting

We have discussed lighting at some length. Lights are often grouped into circuits which may be switched on and off from a central panel in commercial installations, or from wall switches next to doors in residential applications. When there are two or more doors into a room, there is often a light switch placed at each door. Such switches are called *three-way* switches, in which either switch controls the lighting circuit. Thus, the on position of a switch may be either up or down, depending on the position of the other switch. When more than two switches are necessary, two of the switches must be three-way switches and the remaining additional switches must be *four-way* switches.

Motors

Only four types of motors are in general use. The *DC motor* is used for small scale applications and for elevators, where continuous and smooth acceleration to a high speed is important. *Single-phase AC motors* are used in many sizes and shapes, typically 3/4 horsepower or less. Larger motors are typically *three-phase induction motors.* Such motors remain constant in rpm, unless heavily overloaded. Typical power factors are in the range of 0.7 to 0.9. They are characterized by extreme reliability. The only other motor commonly available is the universal motor, which runs on either DC or AC current, but which varies in speed based on the load. These are often found in mixers, hand drills and similar appliances.

All motors should be protected against overload by *thermal relays* which shut off the power when any part of the motor or housing gets too hot.

Capacitors

The simplest capacitor is a set of two plates separated by a small insulating layer. Current is "stored" on one plate and at some point, the entire stored amount is discharged. Capacitors are used to improve the power factor in a circuit. This improves efficiency and overall performance.

Receptacles

A *receptacle* is commonly known as an *outlet,* or erroneously known as a wall plug. (A plug is actually what goes into the outlet.) In residential construction, outlets should be no further than 12 feet apart. All outlets should be three-prong outlets, where the third prong is grounded. All of the outlets in a large room should not be on the same circuit. Thus, when a fuse or circuit breaker trips because of an overload, the room will not be plunged completely into darkness.

Panelboards

A panelboard is a set of fuses or circuit breakers which control the circuit loading in a building. It provides a central distributing point for branch circuits for a building, a floor, or a part of a floor. Each breaker serves a single circuit, and the overload protection is based on the size and current-carrying capacity of the wiring in that circuit. A building may have several panelboards and a main panel, with a disconnect switch for the entire building.

Wiring

The sizes of electrical wiring are standardized using *American Wire Gage (AWG).* No sizes smaller than *14 gage* should be used for building wiring.

Aluminum wiring has been discontinued for small gages (#4 or less) in most areas because of some questions about its oxidation, connection deterioration, and metal fatigue over a long period of time. Several major fires have been

A. ONE SWITCH LOCATION

B. TWO SWITCH LOCATIONS

C. THREE OR MORE
   SWITCH LOCATIONS

LIGHTING SWITCH CIRCUITS

blamed on the use of aluminum, and copper wire is again the standard for branch circuits.

Some circuits must be oversized. For example, the rating of a motor should not be greater than 80% of the capacity of the circuit which feeds it. Thus, any circuit which carries motors should have the rated load multiplied by 1.25 (1/0.80), to determine the wire size. The same factor should be applied to any circuit which is expected to operate continuously for three hours or more.

Conduit

Wires must be physically protected in addition to being insulated. In commercial applications this is accomplished by housing them in a conduit. The conduit size is designated by its interior diameter, and the number of wires of any given size which can fit into a given conduit is specified by code.

There are several types or classes of conduit.

Rigid conduit is the safest conduit, and has the same wall thicknesses as Schedule 40 plumbing pipe. All connections are rigid and threaded, similar to plumbing pipe. The rigid conduit is installed, and then any number of wires up to its rated capacity may be pulled through it. In exterior applications, the conduit must be galvanized, while for interior situations it may simply be enamel coated.

Intermediate metallic conduit (IMC) is a steel conduit with thinner walls than plumbing pipe, slightly less expensive, and generally as acceptable for conduit as rigid conduit.

Electrical metallic tubing (EMT) is the thinnest of the simple metal conduits. It is galvanized, and connections are made with a special clamping system rather than threading. It is sometimes known as *thin wall.*

TABLE 5.2—WIRE GAGES AND CONDUIT SIZES

AWG Size	Ampere Capacity 60 C Insulation	Conduit Size for Three Wires (Inches)
14	15	1/2
12	20	1/2
10	30	1/2
8	40	3/4
6	55	1
4	70	1
3	80	1 1/4
2	95	1 1/4
1	110	1 1/4
0	125	1 1/2
00	145	1 1/2
000	165	2
0000	195	2
250 MCM	215	2 1/2
300 MCM	240	2 1/2
350 MCM	260	2 1/2
400 MCM	280	3
500 MCM	320	3

Flexible metal conduit is available both with and without a flexible waterproof jacket. It is commonly called "flex" or occasionally referred to by the product name "Greenfield." It can be used everywhere except underground.

A conduit similar to "flex" is *interlocked armored cable,* which consists of a prewrapped set of wires encased in an interlocking metal spiral armor. It is factory assembled, and no wires may be added in the field. It is designated *BX* cable, and it may not be used underground or embedded in concrete. No wires may be pulled through it in the field or after it is installed.

In office construction, where there may be several types of cable and power services, and the office layouts may change, there are special *power grid* floors or *cellular metal* floors. The concrete floor is poured directly over the floor system, which has knockout panels at regular intervals to allow access to the different raceways.

An alternative to conduit in residential construction is *sheathed* wire or "Romex." This consists of two insulated live wires and one ground wire, all encased in a plastic sheath. It is officially designated as type *NM* or *NMC* cable. It is strung inside walls, and sometimes in exposed areas such as garages. No other covering is required. It may not be used in commercial garages, and cannot be embedded in concrete.

CALCULATIONS

Most calculations are beyond the scope of this course, and not required of the architect or candidate.

However, you should be aware that the voltage drop due to the resistances of the wire in a given circuit may become noticeable in a large circuit. No more than 3 percent is allowed in lighting circuits, and no more than 5 percent in circuits supporting motors.

Load Estimation

It may be necessary for the architect to estimate the overall electrical load early in a project. This can be done by estimating the wattage per square foot, based on general experience for various building functions.

There are also minimum wattage per square foot requirements to provide for lighting in buildings. Of course, the actual load should be

TABLE 5.3—LOAD ESTIMATING FACTORS			
	Watts/Square Foot		
Load	Low	Avg.	High
Lighting	2	3	5
Convenience Outlets	1	2	3
HVAC*	4	5.5	7
Miscellaneous	0.2	0.5	0.7
*Fuel-fired heating, electrically driven refrigeration.			

determined by the lighting calculations, not by estimates of wattage per square foot. (See Lesson Four.)

SAFETY CONSIDERATIONS

Short Circuits

Strictly speaking, a short circuit occurs when two conductors which are adjacent to each other lose so much insulation that a current flows directly between them. Since there is little resistance, a very high current may result, which can cause the wiring to get quite hot. Combustion within the walls may take place. This is especially dangerous because it may smolder and spread for some time before it is detected.

In practice, the term short circuit is applied to almost any situation where the current is flowing where it shouldn't. Three types of protection are available.

Shutoff Devices

Fuses are devices which are composed of a soft metal link in a glass plug or fiber cartridge, which are rated at a certain current flow. If the current exceeds that rate, the metal link will get hot enough to melt, breaking the circuit. Obviously, fuses are used only once, and must subsequently be replaced. The largest glass plug fuse is rated at 30 amps, but cartridge fuses are available at much higher ratings.

Circuit breakers are devices which automatically disconnect a circuit when the current is excessive. The circuit breaker, however, may be reset after the trouble has been found and corrected. Circuit breakers may be used as a backup switch to shut off an area or circuit being worked on or examined. Although they are more expensive than fuses, circuit breakers are employed in nearly all commercial applications, since they are simple, require no replacement, and therefore have low maintenance costs.

The third type of protection is called a *ground fault (circuit) interrupter (GFI or GFCI).* It detects a continual current lost to ground, even after the power is shut off. This current might not be great enough to cause a fire, and thus might not trip the circuit breaker or melt the fuse, but it is undesirable nonetheless. After detecting such a current, the GFI breaks the circuit. Such devices are required on any circuit of 15 or 20 amps which serves a bathroom, garage, or outdoor area, as well as temporary circuits on construction sites. All large high voltage circuits, such as 480/277 volt 1,000 amps, are required to have ground fault protection.

Grounding

Grounding is a basic safety precaution. A ground wire is fastened to an element which provides a path directly to the ground, thus dissipating any electric current with little or no resistance, and averting possible damage or injury. Thus, many appliances are housed in a metal casing, and the metal casing is grounded. If there is a short circuit inside the case, the current will pass through the case to the ground wire and be harmlessly dissipated rather than

passing through the case to an individual who might touch it, resulting in possible injury.

Ground wires are covered with green insulation, or may even be bare. They are typically fastened at some point to a steel cold water pipe in the plumbing system, which should provide a direct path into and under the ground, where the current will be dissipated. Three-pronged outlets have the third prong connected to a ground wire.

SERVICES

All of the services arriving on the site are called the *service drop,* and consist of the wires from the main line, a transformer, a meter and a disconnect switch.

To avoid excessive voltage drop and flicker, the distance from the transformer to the meter should not exceed 150 feet. The minimum service for a residence is now *100 amps.* In residences, the panel and disconnect are usually located outside the building where they are accessible to firefighters. In commercial construction, they may be located inside the building, but must be directly accessible from an exterior door.

Meters

The electric usage in a building may be measured in two ways. In residential applications, only the total consumption is measured. The unit of measure is the watt hour, which again is usually in thousands, called kilowatt hours (kwh). Costs range from 8 to 18 cents per kwh.

In larger buildings, not only is the total consumption rate measured, but the *peak demand* as well. This is because large peaks require the utility company to build more power generating capacity to meet the peak, which will then sit there, idle, for the remainder of the time. This is extremely inefficient and expensive. The charge associated with the peak demand is called a *demand surcharge*.

Emergency Power Sources

Emergency power is required for lighting exit passages and exit signs, as discussed in Lesson Four. In addition, many institutions such as hospitals require life support equipment and operating room equipment to be safe from power interruption. In many cases, elevators and other equipment may also require backup power.

For lights, backup power is often provided by batteries that are continuously recharged while power is on, and which take over when power is lost. Since batteries are typically 12 volt, fluorescent lamps requires some sort of power conversion.

For larger equipment, a diesel generator with an automatic starting switch and an automatic transfer switch is usually provided in the equipment room. This generator should have at least a *two hour* supply of fuel in reserve.

BUILDING AUTOMATION

Building controls continue to become more and more complex. This is most evident in HVAC and elevator controls, but other building functions are coming under semi-intelligent or computer control. Loads may be shifted to a different time of day to avoid peak demand charges. Lighting may be controlled by time clock or photocell. Fire equipment may be automatically activated and controlled, closing fire doors throughout the building and actuating sprinklers. Emergency intercoms are often provided in high rise construction. Automatic security controls are not only numerous, but sometimes rather creative.

All of this equipment consumes electric power, or is electrically activated, and will continue to increase electrical loading and complexity in buildings as more functions come under computer control, until we begin to have "intelligent" buildings.

SUMMARY

We have learned the basic physics and terms of electricity, and many of the applications in buildings. We have reviewed direct and alternating, single- and three-phase current. We know how to calculate current through series and parallel resistances, and how power is generated. We have been made aware of some of the safety precautions necessary for this amazingly adaptable form of power. Electricity and electronics will continue to affect our lives and our buildings, and understanding the material covered in this lesson will help prepare the candidate for the examination, as well as for actual practice.

LESSON 5 QUIZ

1. Given a 120 volt outlet and a hair dryer with a resistance of 8 ohms (Ω), what is the current flow through the hair dryer when it is turned on?
   - **A.** 15 amps
   - **B.** 6.7 amps
   - **C.** 1.5 amps
   - **D.** .67 amps

2. What is the wattage rating of the hair dryer in the previous question if the power factor is 1?
   - **A.** 1,800 watts
   - **B.** 32 kilowatts
   - **C.** 18 watts
   - **D.** 32 watts

3. Given three *parallel* paths, two paths with a resistance of 4 ohms (Ω) and one with a resistance of 2 ohms (Ω) what is the total net resistance?
   - **A.** .125 ohms
   - **B.** 10 ohms
   - **C.** 1.25 ohms
   - **D.** 1.0 ohms

4. If the voltage between the neutral and the peak of a three-phase current is 120 volts, what is the voltage between two phases?
   - **A.** 120 volts
   - **B.** 208 volts
   - **C.** 240 volts
   - **D.** 277 volts

5. A three-wire transformer means that
   - **A.** the primary winding is one of three input options.
   - **B.** the transformer generates three-phase current.
   - **C.** the secondary winding allows two different voltage takeoffs.
   - **D.** there is a tertiary winding.

6. To provide five light switches for a large room, we need
   - I. two-way switches.
   - II. three-way switches.
   - III. four-way switches.
   - **A.** I and II
   - **B.** II and III
   - **C.** I and III
   - **D.** I, II, and III

7. A variable speed mixer motor is most likely a
   - **A.** universal motor.
   - **B.** DC motor.
   - **C.** three-phase induction motor.
   - **D.** synchronous AC motor.

8. Which conduit system is unacceptable for embedding in concrete?
   - **A.** Rigid conduit
   - **B.** Flex conduit
   - **C.** IMC conduit
   - **D.** Armored cable

9. Which of the following is true about a ground fault interrupter?
   - I. It is a form of protection against short circuits.
   - II. It may be actuated even when there is no equipment switched on in the circuit.
   - III. It can completely disconnect a circuit.
   - **A.** I and II
   - **B.** II and III
   - **C.** I and III
   - **D.** I, II, and III

10. What is a demand surcharge?

    A. A negative charge which builds up in a circuit when the current is too great.

    B. The copper side of a capacitor.

    C. A method by which utilities attempt to reduce the need for new power plants.

    D. The factor that triggers a circuit breaker.

ACOUSTICS

INTRODUCTION

Acoustics is the science of sound. Sound is the sensing, by the ear, of compression waves in a fluid medium, usually air. It is a vital part of our everyday lives, and a strong influence in the design of all buildings. Sound isolation is necessary for privacy, and in many cases for specific tasks, as well. But some buildings also have special acoustical requirements, primarily for musical performances or for communication, which are concerned with clarity and amplification. Therefore, the architect should have a basic understanding of acoustics.

BASIC PHYSICS

In many ways, sound is similar to light. They are both transmitted by waves, and they both obey the inverse square law: the intensity is inversely proportional to the square of the distance from the source. The processes of transmission, reflection and refraction can apply to both sound and light.

Unlike light, however, sound can only be transmitted through a medium, such as air. The velocity of sound through air depends on the barometric pressure and altitude. It can also be transmitted through water, through the ground, or through building structure and materials. It can be reflected off of surfaces, and even focused toward a point by the proper reflective shape. Sound may also be refracted (bent) around objects. Light tends to cast much sharper shadows than sound. A wall in an open field will cast a very sharp light shadow, but its sound shadow is indistinct; sounds from the other side of a free standing wall are often clearly audible.

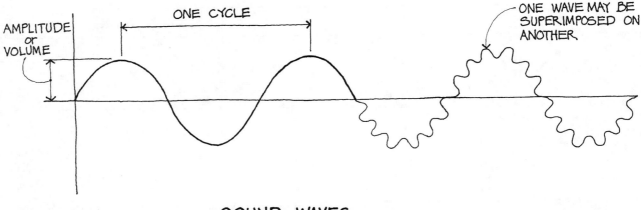

SOUND WAVES

Just as we perceive the wavelength of light in terms of its color, we perceive the wavelength of sound in terms of its *pitch*. Each note of the musical scale represents a specific wavelength or frequency of sound. This is most easily seen when we look at a graph of the sound compression wave in air.

One complete wave form is called a cycle. The frequency of the sound (the pitch) is the number of cycles occurring per second. This may be abbreviated cps, but is more commonly known by the term *Hertz*. Thus, 60 cps = 60 Hz.

A sine wave is a "pure" tone, such as that produced by an electronic instrument. A sound may consist of one wave form superimposed on another. The ear can distinguish several notes at once, for example a chord, or even an entire symphony orchestra playing a single chord.

The ear can distinguish between a square wave at a particular pitch and a sine wave. The square wave sounds harsh, like a buzzer, or even an outboard motor. Different instruments, different voices, and other individually recognizable sound sources have different wave forms.

Although the human ear can hear sounds in the 20 Hz to 20,000 Hz range, it is most sensitive to sounds in the *125 Hz to 6,000 Hz* range. Many animals can hear much higher frequencies. Sounds below 20 Hz are often sensed as vibrations rather than sounds. Structure-borne sound in buildings is sometimes barely audible, but can still cause discomfort.

The height of the wave form is related to the amplitude or magnitude or intensity of the sound. Loud sounds have great amplitude, and represent a larger amount of energy stored in the wave, which may be measured in *watts/cm²*.

The human ear can respond to a tremendous variation in sound amplitudes without being damaged. The ratio between the amplitude of the quietest sound the human ear can hear and that of the loudest sound which does not cause pain is 1:10,000,000,000.

Logarithmic Scales

In order to deal with these variations in a manageable manner, acoustics uses logarithmic scales. The decimal logarithm (log) of a number is the exponent to which the number 10 must be raised to equal that number. For example, the log of 100 is 2, since $100 = 10^2$.

The five basic rules of logarithms are:

$$\log 10^n = n$$

$$\log A \times B = \log A + \log B$$

$$\log A/B = \log A - \log B$$

$$\log C^n = n \log C$$

$$\log 1 = 0$$

Sound Intensity Level

The basic unit of sound intensity level is called the *decibel* (after Alexander Graham Bell) and is expressed by the formula:

$$IL = 10 \log (I/I_0)$$

where

IL = intensity level expressed in decibels (dB)
I = intensity of the sound being measured
I_0 = reference intensity of 10^{-16} W/cm^2, which is the quietest sound which we can hear

Note that the *intensity* of sound is measured in power (watts) per square centimeter, but we generally deal with the *intensity level (IL)*, which is measured in decibels.

To get an idea of how useful the logarithmic scale is, let's solve for the intensity level of a sound which is 10,000,000,000 times as loud as the quietest sound we can hear.

$$IL = 10 \log (I/I_0)$$

$$= 10 \log \frac{10,000,000,000 \times 10^{-16}}{1 \times 10^{-16}}$$
$$= 10 \log (10,000,000,000/1)$$
$$= 10 \log 10^{10}$$
$$= 10 \times 10$$
$$= 100 \ dB$$

Obviously, 100 dB is a great deal easier to express than $10,000,000,000 \times 10^{-16}$ watts/cm^2.

How does this work for small ratios, for example a sound which is twice the intensity of the reference intensity?

$$IL = 10 \log (I/I_0)$$
$$= 10 \log (2/1)$$
$$= 10 \times 0.301$$
$$= 3.01 \ dB$$

The number is still meaningful and easy to express.

Sound Power Level

We can also measure the power *at the source* and convert it to a logarithmic scale. This would measure watts instead of watts/cm^2 or watts/meter^2. The result would be:

$$PWL = 10 \log W/W_0$$

where

PWL = sound power level
W = power at the source measured in watts
W_0 = reference wattage, 10^{-12} watts

Sound Pressure Level

There is a third measure of sound, which is the pressure exerted by the sound wave on a surface at a given location. This is similar to intensity level, but not exactly, since it varies with barometric pressure as well. It is also expressed in logarithmic form:

$$SPL = 20 \log P/P_0$$

where

SPL = sound pressure level
P = pressure at the measured point, in newtons/meter^2
P_0 = reference pressure, 2×10^{-5} N/m^2

Of the three different types of sound measurement, IL is by far the most widely used. Notice the factor of 20 in the equation for SPL, which

TABLE 6.1—TYPICAL IL LEVELS		
Sound	**SPL dB**	**Intensity Watts/Meter^2**
Threshold of pain	130	10^1
Hard rock band	120	10^0
75 piece orchestra	110	10^{-1}
Loud auto horn 10 feet away	100	10^{-2}
Noisy urban street	90	10^{-3}
Truck passing by	80	10^{-4}
Automobile passing by	70	10^{-5}
Conversation	60	10^{-6}
Average office	50	10^{-7}
Quiet office	40	10^{-8}
Unoccupied office	30	10^{-9}
Whisper	20	10^{-10}
Rustle of leaves	10	10^{-11}
Threshold of acute hearing	0	10^{-12}

makes the numerical value of SPL approximately equal to that of IL. Thus, for the purposes of the exam, IL and SPL may be assumed to have the same value. PWL is different, and always represents the power *at the source.*

The problem with using a logarithmic scale is that when we have two sound sources, we cannot simply add their dB levels together. For example, two sources at 60 dB each result in a total IL of 63 dB, *not* 120 dB. Similarly, although doubling the distance from the source to the receiver cuts the intensity to one quarter (remember the inverse square law) this results in a change of only 6 dB in intensity level. Table 6.2 may be used to determine the decibel level which results when two sound sources are added together.

It is always best to use the logarithmic equations, which is not difficult once the five basic rules are understood, as shown in the following example:

Example #1

Given a source of sound ten feet away from the receiver, and an IL of 90 dB at the measured position (receiver),

1. What is the sound *intensity* at the receiver position?

2. What is the *sound power* of the *source*?

3. What is the sound *intensity level* at 80 feet from the source?

Solution:

1. The intensity level at the receiver is 90 dB. We can determine the intensity using the logarithmic formula

$$IL = 10 \log I/I_0 = 10 \log I/10^{-16}$$

$$90 = 10 \log I/10^{-16}$$

$$9 = \log I/10^{-16} = \log I - \log 10^{-16}$$

$$\log I = \log 10^{-16} + 9$$

$$I = 10^{-16} \times 10^9 = \mathit{10^{-7}\ watts/cm^2}$$

TABLE 6.2—ADDING TWO DB LEVELS	
Difference in dB Level	Add to Larger dB Level
0	3.0
1	2.5
2	2.1
3	1.8
4	1.5
5	1.2
6	1.0
7	0.8
8	0.6
9	0.5
10	0.4

2. The sound power of the source is related to the intensity at a given point by the equation

$$I = W/(4\pi d^2 930)$$

where

   I = intensity in W/cm^2

   W = power at the source in watts

   d = distance to the source in feet

   930 = conversion factor to translate between feet and $watts/cm^2$

Transposing that equation to solve for W in terms of I, $W = I \times (4\pi d^2 930)$

Substitute 10^{-7} watts for I, and 10 feet for d:

$W = 10^{-7} \times (4\pi 10^2 930)$
   $= 0.12$ watts

3. Doubling the distance from the source reduces the intensity to one quarter (the inverse square law) and reduces the IL by 6 dB. Since 80 feet is 8 times the distance of 10 feet, the IL may be calculated by successive reductions of 6 dB for each doubling of distance.

Thus, the IL at 20 feet is 6 dB less than at 10 feet, the IL at 40 feet is 6 dB less than at 20 feet, and

the IL at 80 ft is 6 dB less than at 40 ft,

resulting in a total drop of 18 dB at 80 feet, or 90 dB − 18 dB = *72 dB*.

Alternatively, we could use the inverse square law, followed by the logarithmic conversion.

$$I_2 = I_1 (d_1/d_2)^2$$

$I_{80} = I_{10}(10/80)^2 = I_{10} \times 1/64$
   $= (10^{-7} \text{ watts/cm}^2)64$
   $= 1.5625 \times 10^{-9} \text{ watts/cm}^2$

$IL = 10 \log (1.5625 \times 10^{-9}/10^{-16})$
   $= 10 \log 1.5625 \times 10^7$
   $= 10 \times (7.19)$
   $= 71.9 \text{ dB}$

This is essentially the same result as we obtained before.

Weighted Scales for the Human Ear

The human ear is more sensitive to sounds in the middle frequencies than to those in very high or low frequencies. Therefore, weighted measuring scales have been developed. The scale which most closely represents the response of the human ear is called the A scale. When measurements using the A scale are converted to decibels, the resultant measure is designated dBA.

The human ear may also be temporarily affected by exposure to continuous loud noise levels. There can be as much as a 30 dB loss in sensitivity. Permanent damage is also possible. For this reason, the Occupational Safety and Health Administration (OSHA) has developed requirements intended to limit exposure to high noise

TABLE 6.3—OSHA PERMISSIBLE NOISE EXPOSURE	
Duration per day (in hours)	**Sound Level, dBA (slow response)**
8	90
6	92
4	95
3	97
2	100
1 1/2	102
1	105
1/2	110
less than 1/4	115

levels in places of employment. Table 6.3 presents the permissible exposure limits for continuous sound levels.

As with vision, the mind integrates the incoming sensory information from the ear. If one ear hears something a few thousandths of a second before the other, the mind infers the approximate direction of the source. The exception is when a source is either directly in front of, above, or directly behind the head, since the sound arrives at both ears simultaneously, and direction cannot be determined.

TRANSMISSION AND REFLECTION

All of the methods above refer to the behavior of sound in a free environment, such as an open field. However, there are some considerable distortions when we get into buildings. Just as with light, there is a great deal of acoustical reflection inside of a room. Near the source, the inverse square relationship holds, but as we get further from the source, the reflections dominate, and the distance relationship becomes less important, or even unimportant.

Similarly, one of our concerns within buildings is the transmission of sound from one room into another room through the separating wall. These two phenomena are both affected by sound absorption.

Sound Absorption

The reflection of sound in a room causes two things to occur. The noise level, or volume of

TABLE 6.4—TYPICAL SOUND ABSORPTION COEFFICIENTS						
	125 Hz	**250 Hz**	**500 Hz**	**1,000 Hz**	**2,000 Hz**	**4,000 Hz**
Heavy carpet on concrete	.02	.06	.14	.37	.60	.65
Coarse concrete block	.36	.44	.31	.29	.39	.25
Heavy velour drape	.14	.35	.55	.72	.70	.65
Plate glass	.18	.06	.04	.03	.02	.02
1/2" gypsum board on 2 x 4 studs	.29	.10	.05	.04	.07	.09
Smooth finish plaster on lath	.14	.10	.06	.04	.04	.03
Water surface, at rest	.008	.008	.013	.015	.02	.025
Audience in upholstered seats, per square foot of seating area	.60	.74	.88	.96	.93	.85

sound, is greater than it would be for the same source in an empty field. There is a delay factor as well, so some sounds persist for a time in a very reflective space, even after the source has stopped. This is known as reverberation. Although reverberation is similar to an echo, in some respects it is different.

An echo is a discrete reflection of a sound, usually delayed 1/10th of a second or more. With a sufficient delay, an entire word may return separate and intact. A prime example is the echo from a hard canyon wall, or some other distant, fairly flat object. Reverberation is a more continuous reflection, over shorter time spans, such as an organ note dying out slowly in a cathedral, as the sound reverberates back and forth. Not only is sound magnitude related to absorption, but so is reverberation time.

There is an acoustical measure of reflectivity and absorptivity similar to that in radiation, designated by the Greek letter alpha (α). It is measured in a unit called a sabin, after Wallace Clement Sabin, a physicist who pioneered acoustical work.

Sabin lectured in a physics auditorium in which most lectures were unintelligible. He determined that this was because sounds took so long to die away that the syllables of one word would run together and even into the next word. He was able to reduce this phenomenon by putting horsehair pillows on the seats. The pillows absorbed the sound instead of reflecting it back, and there was a vast improvement in the students' ability to understand his lectures.

The absorptivity per square foot of any given surface varies from 0 (all sound is reflected) to 1.0 sabin (all sound is absorbed).

The absorptivity of a room is the sum of the different surface areas times their respective absorptivities, and is expressed by the equation:

$$A = S_1\alpha_1 + S_2\alpha_2 + S_3\alpha_3 + ... + S_n\alpha_n$$

where
   A = total absorptivity
   S_1 = surface area of material 1 in square feet
   α_1 = absorptivity of material 1 in sabins
   S_2 = surface area of material 2 in square feet
   α_2 = absorptivity of material 2 in sabins

and so forth, for all the materials present in the room.

Materials may have different absorptivities at different frequencies, so a complete table will include the α value for each frequency and an overall *noise reduction coefficient (NRC)*, which is the arithmetic average of all of the different frequencies from 250 Hz to 2,000 Hz.

Example #2

Find the total absorptivity of an empty room at 1,000 Hz. The room is 20 × 30 feet in plan and 10 feet high. The floor is concrete covered with carpet, one 30 foot wall is plate glass, and the other three walls are half inch gypsum board over studs. (It is not unusual to disregard doors when calculating absorptivity, but they are most important when dealing with transmission.) The ceiling is of smooth finished plaster on lath.

Solution:

$$A = 2 \times 20 \times 10 \times .04 \text{ (two 20 foot walls)}$$
$$+ 1 \times 30 \times 10 \times .04 \text{ (30 foot wall)}$$
$$+ 1 \times 30 \times 10 \times .03 \text{ (30 foot window)}$$
$$+ 1 \times 30 \times 20 \times .37 \text{ (carpet on floor)}$$
$$+ 1 \times 30 \times 20 \times .04 \text{ (plaster ceiling)}$$
$$= 16 + 12 + 9 + 222 + 24$$
$$= 283 \text{ sabins}$$

most of which is represented by the carpet!

We can calculate the volume (amplitude) of the sound in an enclosed reverberant space using several equations. If we know the power in watts at the source we can calculate the intensity (watts/meter^2) from the equation

$$I = P/930A$$

where
  I = intensity throughout the space (except very close to the source), in watts/meter^2
  P = power at the source, in watts
  A = total absorptivity of the space, in sabins

We can find the intensity level (IL) using the logarithmic equation on page 107. Once we know the IL in a room for a given condition, we can find the IL resulting from any given change from an easier formula:

$$NR = 10 \log A_2/A_1$$

where
  NR = reduction in sound from case 1 to case 2, in decibels
  A$_2$ = total absorptivity of the room in case 2, expressed in sabins
  A$_1$ = total absorptivity of the room in case 1, expressed in sabins

Example #3

What would be the absorptivity if we provided drapes across the window in Example #2? What would be the reduction in sound?

Solution:

    A = 2 × 20 × 10 × .04 (two 20 foot walls)
      + 1 × 30 × 10 × .04 (30 foot wall)
      + 1 × 30 × 10 × .72 (30 foot drapes)
      + 1 × 30 × 20 × .37 (carpet on floor)
      + 1 × 30 × 20 × .04 (plaster ceiling)
      = 16 + 12 + 216 + 222 + 24
      = *490 sabins*

compared to 283 sabins.

The reduction in sound in that room would be

$$\begin{aligned} NR &= 10 \log A_2/A_1 = 10 \log 490/283 \\ &= 10 \log 1.73 \\ &= 10 \, (.238) \\ &= 2.38 \ (dB) \end{aligned}$$

We could still include the absorptivity of the glass wall, since it hasn't really been removed. It is not a significant proportion of the total. Hanging fabric tapestries or drapes in a concert hall provides one way of varying the absorptivity of the hall for different functions. The drapes may be retracted, or the tapestries removed, as necessary.

Reverberation

As we previously discussed, reverberation is the slow fading of a sound in an enclosed space. The amount of time which elapses before there is complete silence after a 60 dB sound has stopped is called the reverberation time (T$_R$). As spaces become larger, the reverberation time tends to increase. As absorptivities become greater, the reverberation time tends to decrease. This relationship is expressed by the formula:

$$T_R = .049 \ V/A$$

where
  T$_R$ = reverberation time in seconds
  V = volume of the space in cubic feet
  A = total absorptivity in sabins

Different functions have different optimum reverberation times. In Table 6.5, the optimum reverberation times for various functions are displayed in bar chart form.

Typically, speech should have a short reverberation time, to avoid the overlapping of syllables and resultant unintelligibility. Conversely, a pipe organ sounds best when there is a longer reverberation time.

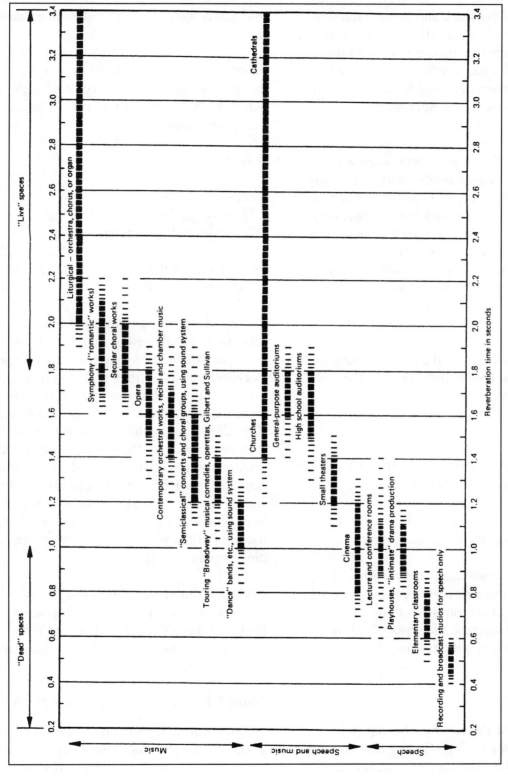

TABLE 6.5 — REVERBERATION TIMES

Reprinted courtesy of BBN Technologies.

As noted above, the reverberation time depends on the volume of the space and the absorptivity. Since the volume is generally determined by other factors, such as audience size, sight lines, building size, etc., the architect usually obtains the desired reverberation time by using materials which will provide the appropriate absorptivity.

Note that the audience itself represents a very large absorptive surface area, so a theater which is reverberant when empty becomes much less so when occupied. Using absorptive upholstery on the seats reduces the variation with audience size. Very reverberant spaces are often called "*live*" spaces, and spaces with short reverberation times are called "*dead*" spaces.

Example #4

What is the reverberation time of the room we calculated above, both before and after the addition of the drapes?

Solution:

$$T_R = .049\,V/A = .049 \times 20 \times 30 \times 10/283$$
$$= 1.04 \text{ seconds (before drapes)}$$

$$T_R = .049\,V/A = .049 \times 20 \times 30 \times 10/490$$
$$= 0.60 \text{ seconds (after drapes)}$$

If we were concerned about the reverberation time being a bit too long for speeches, we could provide acoustical tile on the ceiling, or add more drapes, or increase the absorptivity in some other way without changing the room size.

Room Acoustics

An auditorium or theater must be carefully designed to produce a satisfactory acoustical environment. The floor area is determined by

the number of seats to be provided. As a general rule, good sight lines will also give good sound lines.

After selecting a reverberation time T_R to suit the purpose of the hall, the ceiling height of the hall must be established. A rule of thumb is to make the average ceiling height

$$H = 20 \times T_R$$

where
   H = height in feet
   T_R = desired reverberation time

The volume of the hall should be at least 100 cubic feet per person.

The two basic design goals are to reinforce reflections which arrive at the listener at nearly the same time as the sound from the source, and to cancel out reflections that would be excessively delayed. Thus the stage itself may have reflective surfaces, and the rear of the auditorium almost always has an absorptive surface or a shape which traps the sound. Sloping both the seating and the ceiling upward away from the stage is beneficial in two ways. The audience sight line is improved because the rear seat rows have visibility over those in front of them, and listeners in the rear can also receive sound by direct path from the stage.

Reflected sound from the ceiling is useful, but where speech intelligibility is important, the length of the reflected path should not exceed the length of the direct path by more than 34 feet.

Sound Transmission and Isolation

Besides reinforcing or controlling sound within a space, the most important consideration for the architect is limiting the communication of

LOUDSPEAKER

REFLECTIVE CEILING PANELS

REFLECTED PATH

DIRECT PATH

HIGHLY ABSORPTIVE TREATMENT

ELEVATION

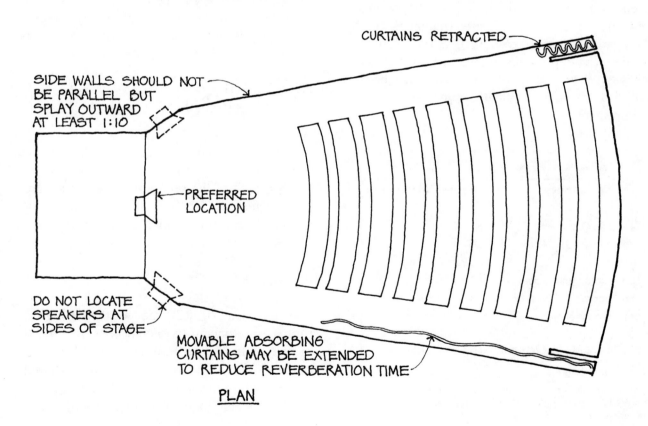

CURTAINS RETRACTED

SIDE WALLS SHOULD NOT BE PARALLEL BUT SPLAY OUTWARD AT LEAST 1:10

PREFERRED LOCATION

DO NOT LOCATE SPEAKERS AT SIDES OF STAGE

MOVABLE ABSORBING CURTAINS MAY BE EXTENDED TO REDUCE REVERBERATION TIME

PLAN

AUDITORIUM DESIGN FEATURES

sound between two spaces. A certain level of quiet is required for the performance of many tasks, and excessive noise usually interferes with communication.

A family of curves, known as the 1957 Noise Criteria or NC curves, are widely used in specifying the maximum noise level in a given space under a given set of conditions. These are shown on the following page. If a given NC curve is specified, then the noise level in each octave band centered around the frequencies shown must not exceed the SPL level intercepted by the specified curve. For example, if the NC 30 curve is specified, the 250 Hz octave band must not exceed 41 dB as measured on a sound meter. To verify that the specified NC curve has been satisfied, octave band SPL readings must be made at each of the indicated frequencies and the results compared to the NC curve. If all of the measured values are equal to or below the intercepts of the curve, the curve has been satisfied. Table 6.6 indicates some typical NC requirements.

TABLE 6.6—SUGGESTED NC CURVES		
Occupancy	Minimum	Maximum
Broadcast studio, concert hall	NC-15	NC-20
Bedrooms	NC-20	NC-30
Large conference room, courtroom	NC-20	NC-30
Small conference room, library, private office	NC-30	NC-35
General office area, stores, cafeterias	NC-35	NC-40
Minimum values provide sufficient background noise for masking, maximum values prevent outside interference.		

A similar set of curves, known as the 1971 Perceived Noise Criteria (PNC) has been proposed, which is used in much the same way as the NC curves.

Noise Reduction Through a Wall

An ideal wall separating two non-reverberant spaces would transmit some fraction (called τ) of the sound incident upon it (hitting it) in the source room to the other space. *Transmission loss (TL)* is defined as

$$TL = 10 \log 1/\tau$$

Thus, as τ increases, TL decreases.

In reverberant rooms the difference in IL between two real rooms separated by a real barrier is termed *Noise Reduction (NR)* and is related to the ideal TL by the equation:

$$NR = IL_1 - IL_2$$
$$= TL - 10 \log S/A_R$$

where
   TL = the free field transmission loss of the wall
   S = area of the separating wall, in square feet
   A_R = total absorptivity in the receiving room, in sabins

Example #5

Two identical 20 × 30 foot × 10 foot high rooms are adjacent to one another, and the IL in one room is 75 dB. What is the IL in the other room if they are separated by a shared 20 foot wall with a TL of 25 dB and the previously calculated absorptivity of the room is 490 sabins?

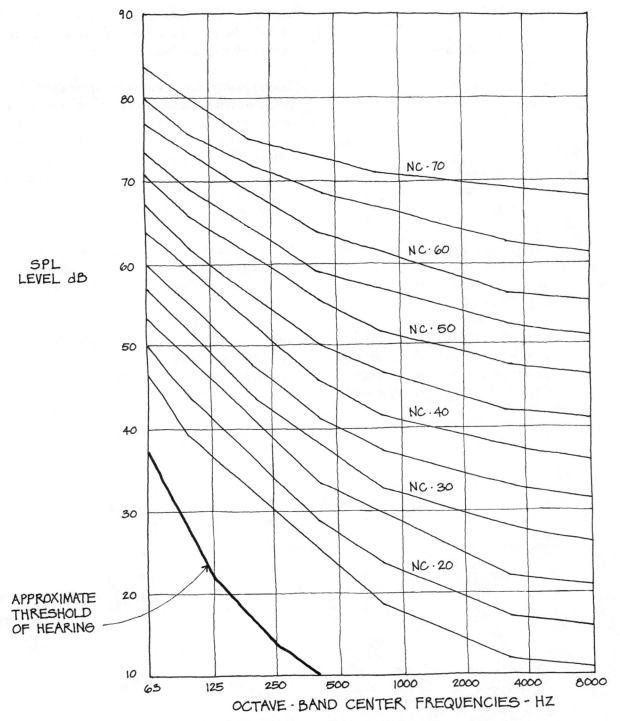

NOISE CRITERIA CURVES - 1957 NC

Solution:

$$NR = TL - 10 \log S/A_R$$
$$= 25 - 10 \log 20 \times 10/490$$
$$= 25 - 10 \log .408$$
$$= 25 - [10 \times (-.389)]$$
$$= 25 + 3.89$$
$$= 28.89 \; dB$$

$$IL_2 = IL_1 - NR$$
$$= 75 \; dB - 28.9 \; dB$$
$$= 46.1 \; dB$$

The TL of a wall is determined by its construction, its stiffness, and its mass. A general rule of thumb is that for every doubling of mass there is an increase of between 5 and 6 dB in TL.

Thus, one way to reduce the sound transmission of a wall is to increase its mass. For example, a concrete wall has much less sound transmission than a stud wall. However, concrete walls tend to be uneconomical.

Other ways of improving the TL of a wall include the use of staggered studs and flexibly mounting gypsum board, as shown below.

Sound always leaks through the weakest point in the isolating wall. Thus a concrete wall with an open window is like a bucket with a hole in it. Almost all of the sound goes through the window. It is of paramount importance, therefore, that holes for pipes, ducts, conduit, etc., be packed with insulation and sealed as well as possible.

It is important to understand the difference between transmission and reflection. Acoustical tile will stop reflection, but will not stop transmission. On the other hand, concrete stops transmission, but does not stop reflection. Do not confuse the processes. Acoustic tile and insulation may stop reflection *inside* a wall cavity, and thus reduce overall transmission. But applying acoustic tile to the outside surface of

GYPSUM BOARD

STAGGERED STUDS

FLEXIBLE METAL CHANNEL

GYPSUM BOARD

BOARD GLUED TO FLEXIBLE CHANNEL

FLEXIBLY MOUNTED GYPSUM BOARD

METHODS OF IMPROVING THE TL RATINGS OF WALLS

a wall does very little to reduce transmission. It reduces *reflection* back into the room, but does not stop *transmission.*

Sound Transmission Class

The *sound transmission class (STC)* is a widely accepted method of rating walls, doors, etc. in terms of their typical or overall resistance to sound transmission. Obviously, the transmission of different sound frequencies varies from wall to wall, so a weighted average of all frequencies is used.

The STC rating of a given wall section is established by measuring the TL of a test panel at 16 one-third octave bands and plotting these as in the following drawing. The standard STC contour is fitted, as closely as possible, to the curve. Then the 500 Hz dB value of the standard curve is used as the STC rating of the test panel. The shape of the standard STC contour was chosen to be representative of a nine inch thick brick wall. Stud walls and other nonconforming walls do not fit as well, and are not as accurately represented by the measurement.

STC values are widely quoted in catalogs and advertising, and are useful for design. The architect should use these values with some caution because actual construction practices tend to be less perfect than those used to construct the test panel.

Impact Noise

Impact noises are erratic sounds caused by footfalls, dropped objects, the vibration of mechanical equipment, etc. The resulting vibration of the structure is then manifested as airborne sound radiated from other locations.

A standardized method of measuring the degree of isolation of impact noise in the structure has been developed. The method utilizes a special "tapping machine" which is placed on

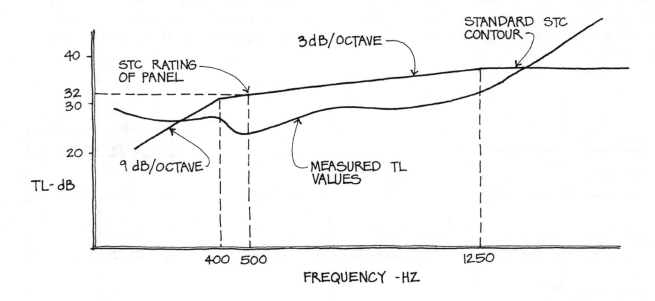

STC DETERMINATION OF A PANEL

the floor to be tested. A sound meter with one-third octave filters is located in the room beneath the tapping machine and is used to measure SPL at each frequency. The resulting data is then plotted, and an *impact isolation class (IIC)* contour is fitted in accordance with prescribed rules. The 500 Hz intercept of the IIC contour becomes the IIC rating of the floor/ceiling assembly. The IIC ratings are utilized similarly to STC values.

IIC values can be improved in several ways. Carpeting and resilient tile floor coverings are effective at middle and high frequencies. Suspended ceilings are also very effective at middle and high frequencies, especially if combined with carpeting. A concrete slab floated on compressed glass fiberboards laid on the structural floor is very effective in improving the performance at all frequencies and, thus, the IIC rating.

Code Requirements

In addition to OSHA requirements for noise levels at job sites, the Uniform Building Code has established minimum airborne and impact sound isolation values for residential occupancies. Dwelling units and guest rooms must have some acoustical privacy between each other and between them and public areas, such as corridors and service areas. General guidelines are STC 50 for walls, floors and ceilings, and STC 26 for entrance doors.

Speech Privacy

Speech privacy is an important factor of concern in offices, apartment houses, etc. Conversations overheard from the other side of a partition are annoying, and also cause concern that one's own conversations may be overheard beyond the walls. One of the factors that determines whether overheard speech is intelligible is the background level of noise. This gives the designer an additional tool to employ in creating privacy: increasing the background noise level in an innocuous way. This is often called masking noise or white noise. The rushing sound of air from HVAC outlets can provide such sound, because it represents a good mix of all frequencies. Do not confuse this with the knocks, squeaks and vibrations sometimes

TYP. IIC = 25
POOR

TYP. IIC = 47
GOOD

TYP. IIC = 60
BEST

METHODS OF IMPROVING THE IIC RATINGS OF FLOORS AND CEILINGS

POOR BARRIER LOCATION

GOOD BARRIER LOCATION

SOURCE

RECEIVER

OUTDOOR SOUND BARRIER

present in mechanical equipment, which tends to mask little, and is a source of annoyance.

SPECIAL CASES

Outdoor Sound Barriers

There is frequently a need to control outdoor sounds, as from a highway near a structure. Solid barriers, as shown above, can be effective, but they must be properly located. The best location for a barrier is very close to the source or very close to the receiver; middle positions are significantly less effective. The barrier must be higher than the line-of-sight between the source and the receiver; the higher the better. Unfortunately, the attenuation of a barrier is not constant, but instead increases at 3 dB per octave. For this reason, low-pitched noises, such as truck engine sounds, are not reduced as much as tire whine, which is at a higher frequency.

Attenuation values of 10 dB or more at 500 Hz are possible; the upper limit is about 25 dB at the higher frequencies.

The use of trees and vegetation as an acoustical barrier was once thought to be effective. However, experiments show that they provide almost no attenuation at all for transmitted

sounds. Vegetation in front of a barrier may reduce the reflected sound, but vegetation alone does very little to stop the transmission of sound.

Mechanical Systems

Control of noise generated by mechanical systems is an important aspect of modern design. Most mechanical equipment produces noise having a definite frequency or pitch related to the rotational speed of the equipment. Of course, numerous harmonics also exist in most cases, producing a larger band of frequencies from any given source. Typical sources are motors, fans, pumps, and compressors. Moving fluids, such as air and water streams, also create noise due to turbulence in the flow, and the noise generated by such flow has no definite frequency; instead, it contains a random mixture of essentially all frequencies. This is what we referred to previously as "*white noise.*"

A low background level of white noise can mask other sounds, particularly in the frequency range of speech. This provides privacy by making neighboring conversations unintelligible when their volume drops below that of the white noise.

The use of high quality mechanical components will help to reduce system noise, but the

ISOLATION HANGERS PROVIDE RESILIENT SUPPORT TO PIPING

SOUND ABSORBING MATERIAL

FAN SHOULD BE DYNAMICALLY BALANCED

FLEXIBLE DUCT CONNECTION

FLEXIBLE LOOPED CONDUIT CONNECTION

MOTOR

MECHANICAL DUCT WITH INTERNAL GLASS FIBER LINING

SLAB PENETRATION PACKED WITH GLASS FIBER AND CAULKED WITH NON-HARDENING SEALANT, BOTH SIDES.

RIGID BASE FOR STABILITY AND TO DISTRIBUTE EQUAL LOADS TO SPRING SUPPORTS
CLEARANCE FOR INSPECTION AND CLEANING

STEEL SPRINGS

RIBBED NEOPRENE PADS TO DAMP OUT HIGH FREQUENCY VIBRATIONS

DESIGN FEATURES TO CONTROL MECHANICAL EQUIPMENT NOISE

designer must still pay attention to details of design to insure satisfactory sound levels in the completed structure. The motor and its load must be rigidly mounted on a common sub-base to insure shaft and/or belt alignment, and the sub-base should be isolated from the structure by springs and pads of rubber, cork or neoprene. The resonant frequency of the sub-base, motor and machine assembly should be one third, or less, of the frequency of the motor. Concrete is sometimes added to increase mass and lower the resonant frequency. Of course the conduit, pipes and duct work must be equipped with flexible connections to avoid transmitting mechanical

vibrations into the remainder of the system. Some of those features are shown in the illustration above.

Long ducts may be acoustically lined to control sound waves through the ducts, but where sound must be controlled in short duct sections, mufflers as short as three feet are available with good attenuation over a wide range of frequencies. Other sound control features include shock arrestors on water pipes to control water hammer and resilient packing to seal around pipes and ducts where they pass through structural openings.

SUMMARY

We have discussed the behavior of sound, which is similar to light, in that it follows the inverse square law, and may be transmitted, absorbed, or reflected. We have also shown examples of calculations using the logarithmic relationships typical of acoustics. We have discussed the sensitivity of the human ear, and code and OSHA regulations designed to protect the ear. We have covered the behavior of sound within an enclosed space and through walls between spaces.

The coverage of acoustics in this lesson should provide the architectural candidate with the basic understanding that he or she is expected to have.

LESSON 6 QUIZ

1. What criteria are used in specifying the maximum noise level in a given space under a given set of conditions?
   - A. PWL
   - B. NRC
   - C. NC
   - D. NR

2. What is the logarithmic measure of the intensity of a sound relative to a reference intensity in watts/m² or watts/cm²?
   - A. IL
   - B. SPL
   - C. STC
   - D. NC

3. Which of the following is a method of rating wall sections or doors according to their typical or overall resistance to sound transmission?
   - A. SPL
   - B. STC
   - C. PWL
   - D. NR

4. Reverberation time is
   - I. the time required for a sound to decay 60 dB in a space.
   - II. the time it takes for an echo to return.
   - III. longer in a dead space, and shorter in a live space.
   - A. I only
   - B. II and III
   - C. I and III
   - D. I, II, and III

5. The maximum acceptable IL for a hospital is
   - A. 100 dB.
   - B. 70 dB.
   - C. 40 dB.
   - D. 10 dB.

6. Doubling the distance from the source to the receiver would result in a
   - A. drop of 6 dB.
   - B. gain of 6 dB.
   - C. drop of 3 dB.
   - D. gain of 3 dB.

7. Doubling the number of sources at a given intensity would result in a
   - A. drop of 6 dB.
   - B. gain of 6 dB.
   - C. drop of 3 dB.
   - D. gain of 3 dB.

8. Doubling the mass of a theoretical wall causes its sound transmission to
   - A. decrease 6 dB.
   - B. increase 6 dB.
   - C. decrease 3 dB.
   - D. increase 3 dB.

9. Given a concrete room with concrete walls, floors, and ceiling, adding carpeting and drapes does which of the following?
   - I. Increases the absorptivity of the room
   - II. Decreases the reverberation time of the room
   - III. Decreases the intensity of the sound that a small motor would make in the room
   - A. I only
   - B. II and III
   - C. I and III
   - D. I, II, and III

10. A partition with a sound transmission class rating of 55 dB would be

    A. very good at stopping reflected sound.

    B. barely acceptable at stopping transmitted sound.

    C. very good at stopping transmitted sound.

    D. barely acceptable at stopping reflected sound.

WATER SUPPLY AND DRAINAGE SYSTEMS

INTRODUCTION

Among the systems which we take most for granted in our buildings are those which provide clean healthy water, and which remove waste water efficiently, without danger, and without breeding disease. Although we give very little thought to modern plumbing, and accept it as a matter of course, obtaining clean water is a major problem in many parts of the world, and its lack is a significant source of disease. The architect should therefore have an understanding of the basic principles of plumbing systems, and the Architect Registration Examination tests this understanding. Let us begin our discussion by separating supply and waste systems.

Supply systems are those which supply clean, clear and potable water for industrial processes, washing, cooking, or drinking. Supply systems are under pressure, and thus must be sealed. Being under pressure, water supply lines can run vertically in buildings, typically in smaller pipes than those used for waste systems. *Sanitary waste* systems, on the other hand, remove contaminated water, and are generally not under pressure. Therefore great care must be taken to drain them by gravity, and to avoid contaminating other systems. *Storm drains* are somewhat similar in that they must be drained by gravity, and typically require much larger pipes. Each of these will be discussed separately, in detail.

SUPPLY

Water must be clean, clear, and *potable*—that is, suitable for drinking. There are several contaminants which may cause trouble and must therefore be considered.

Acidity

When water falls from the sky as rain, it is usually clear of any mineral content, but slightly *acidic*. Acidity is measured by the pH of the water. Neutral water has a pH of 7. The greater the acidity, the *lower* the number. For example, a pH of 6.9 to 6.0 represents slightly acidic water, and a pH of 5 represents very acidic water. A pH *above* 7 represents a *basic* or *alkaline* solution. The pH scale ranges up to 14, which represents the most alkaline, or "basic" solution. The naturally acidic state of rain water has been exacerbated in some areas by the presence of the by-products of combustion in the air, most notably sulfur and nitrogen compounds, which may combine with the moisture present to form sulfuric (most common) or nitric

(less common) acid. This has become a major problem in the northern United States and Canada, where the acidity of many fresh water lakes has risen to the point where few fish, or no fish at all, can survive. This may pose a threat to our water supply at some future date. In any case, the acidity of rain water can cause problems in plumbing systems, since it *corrodes* metal pipes. Furthermore, other contaminants from air pollution may be absorbed into the rain water, which may make it dangerous or unfit for human consumption. For this reason, rain water and *surface runoff* may not be as healthy as water which has been absorbed into the ground and partially filtered of these contaminants.

Hardness

Once water has seeped into the ground, it dissolves minerals present in the ground, most notably *limestone (CaCO_3)* and/or *calcium* and/or *magnesium*. This is called *hard* water. This is not often hazardous for humans, but is harmful to plumbing because it deposits out again, eventually clogging the flow through the pipes. Hardness is not to be confused with acidity—they are two separate problems.

Hard water not only causes deposition in pipes, but is particularly problematic in heat exchangers, such as hot water tanks or hydronic systems. The deposits choke off the flow, or insulate the pipe so that the heat exchange is much reduced. A piece of metal is often inserted in the water tank to divert the deposition rather than allowing it to form on the heat exchanger. This is called an *anode*.

Hardness also interferes with the cleaning and lathering capabilities of detergents and soaps. Since there are already ions dissolved in the water (from the minerals), getting additional dirt to dissolve becomes more difficult. Sometimes the minerals actually coagulate with the soap to

WATER SOFTENER RECHARGE

form a soft paste in the plumbing and a fine film or build-up in whatever is being washed.

Water may be *softened* by removing the mineral ions, or by combining them with something which will not solidify when the water is heated. This is done using the *zeolite* or *ion exchange* process. Such a water softener consists of two tanks. The first contains the zeolite mineral and the second contains salt crystals. The water to be softened is passed through the zeolite tank.

Eventually the zeolite needs to be recharged. It is first backwashed to free clogged material, which is simply drained away. It is then regenerated with brine from the salt crystal tank. The sodium ions from the salt exchange with the magnesium and calcium ions which were trapped by the zeolite, and this is also drained away. Finally, the zeolite is rinsed to remove excess salt, and then returned to normal use.

Carcinogens

Unfortunately, natural minerals are not the only contaminants found in *ground water*. Many other substances have been added to our streams and water supplies, or even to the ground around wells or water supplies, which eventually are passed into the ground water. Most notable are the *carcinogens* (cancer causing agents) such as *PCB's (poly chlorinated biphenyls), DDT* and other insecticides, and *asbestos* fibers. Some areas of the country have experienced major contamination of ground water from the improper disposal of dangerous chemicals.

Consequently, in recent years, our water supply, which we used to trust implicitly, has become more suspect, and as a result, the use of bottled water has increased substantially. But what of the major portions of society who cannot afford bottled water?

Disease

Another type of contamination relates not to dissolved chemicals, but rather to the presence of bacteria or viruses in the water supply. This comes, most often, from the improper disposal of human or animal waste or other organic materials which, in decay, provide the source and breeding ground for disease-causing bacteria and viruses.

The traditional treatment for public water supplies consists of settling out contaminants, sometimes by adding a coagulant such as alum, and letting sediment settle out. *Chlorine* may be added to kill or reduce bacteria. The level is usually about 0.5 ppm (parts per million). If the level exceeds 1 ppm, a distinct chlorine taste is noticeable. *Fluorine* is sometimes added as well, to improve resistance to tooth decay. Oxygen is usually present, but if not, the water is oxygenated by passing it through a spray or a waterfall. The oxygen makes it more fit for human consumption, but also increases the rate at which ferrous (iron based) fittings *rust* or *oxidize*.

MATERIALS AND METHODS

Historically, a number of different materials have been used for plumbing. At one time, lead

was actually used for pipe, primarily for drainage, because it was malleable (bendable by hand) and easy to work with. Lead pipe is now found only in old buildings. Each plumbing material has its own characteristics and typical connections.

Steel

Untreated steel was originally called black iron, because of its color. It was very susceptible to rust and corrosion and was replaced by *galvanized* steel pipe, steel pipe with a thin layer of zinc bonded to the surface to make it comparatively rust-resistant. Wall thicknesses for various pipe diameters are standardized by schedules, of which *Schedule 40* is the most common.

Steel pipe is typically joined mechanically. The two ends to be joined are threaded with a sloping thread, joint compound or tape is applied to seal the minute cracks or gaps, and the two pieces are screwed into a connecting collar with internal threads. When used in drainage systems, the two ends are often clamped together with a rubber sleeve, a steel jacket, and two steel band clamps.

STEEL PIPE CONNECTIONS

Copper

Copper tubing is often used for supply piping, and is considered to be the best material for that purpose. Copper does not rust, and is resistant

COPPER PIPE CONNECTIONS

PLASTIC PIPE CONNECTIONS

to corrosion because oxidation builds up a thin film which protects the copper. The wall thicknesses are much less in copper than in other materials. There are three categories of copper tubing: Type K, Type L, and Type M, with M being the most common and having the thinnest walls.

Copper pipes are joined by a form of *soldering* called *sweating*. *Flux* is applied to clean the surfaces to be bonded, and the sections are heated to a temperature which will melt the flux. The joint is assembled, using a sleeve or elbow which fits closely over the two sections. The solder is applied to the joint, melts, and is drawn into the joint by the capillary attraction between the two surfaces. When the joint cools, the pipes are structurally joined and completely sealed. One advantage of this method is that it is reversible. If the joint is reheated, the solder will melt, allowing the pipes to be slid apart.

Plastic

The prime competitor of copper is plastic pipe. Plastic pipe is available in two types, the PVC pipe, usually used for supply piping (white, with light blue lettering) and the ABS pipe, which is typically used for drainage (larger, black, with white lettering). Both are joined in the same

manner, but using different solvents or cements. Plastic pipe does not corrode, and does not allow the electrolysis which deposits mineral ions. This makes it very durable when properly used. It deteriorates, however, when exposed to ultraviolet light, which is present in direct sunlight. For this reason, plastic pipe should never be used in exposed locations above ground or outside walls. When installing irrigation pipe, for example, all above ground fixtures should be steel or a different plastic or vinyl compound, which is typically green in color.

Plastic connections are made in a manner similar to copper connections. The surfaces are primed, and after the primer has dried, a cement or solvent is applied and the joint is slid together. The connection is structurally complete and sealed. However, the process is not reversible. There is no way to pull the pieces apart if they have been properly joined. Therefore, if changes are necessary, the original connection must be cut out with a saw (not very difficult) and new pipe provided for the gap created.

WATER PRESSURE

As mentioned earlier, water is usually supplied under pressure, which ordinarily should be sufficient to provide a reasonably high flow rate to the upper stories of a building. However, this is not always the case; water is fairly heavy, and lifting it requires a great deal of pressure.

Lift

To understand the relationship between pressure in pounds per square inch (psi) and static head (inches or feet of water which could be supported by a given pressure), one should know that a pressure of one psi is capable of lifting a column of water 2.3 feet high. Thus 10 psi could lift water 23 feet and 100 psi could lift it 230 feet. It is often necessary to have some pressure left at the top of the lift to flush a toilet or other fixture, as shown in the following example.

Example #1

A 10-story building has a floor-to-floor height of 12 feet. A pressure of 15 psi is required to flush a toilet. What is the required water pressure at the base of the building?

Solution:

Total lift = 10 stories × 12 feet/story
= 120 feet

Since 2.3 feet is equivalent to 1 psi, 120 feet is equivalent to 120/2.3 = 52.2 psi.

Required water pressure
= 52.2 psi for lift + 15 psi for flush
= *67.2 psi*

If the pressure at the street level, say the street water main, is less than 67.2 psi, other means must be employed to supply the toilets at the upper floors. These will be discussed shortly.

Note that when converting from pressure (psi) to lift (feet) we use 2.3 feet per psi. When

converting from height or lift (feet) to pressure (psi), we use 1/2.3, or .433 psi per foot.

Different fixture types require different pressures, which vary from 7 or 8 psi min for a faucet to 30 psi min for a hose bibb. Very high pressures cause undue wear on the valve seats and washers. Therefore, if pressures exceed 80 psi, a pressure regulator should be installed, which keeps the pressures in the 40 to 60 psi range.

As mentioned in Example #1, if the pressure in the water main is insufficient to supply the upper stories of a building, special systems must be employed. These include the downfeed system, the pneumatic tank system, and the tankless system.

In the *downfeed* system, a tank mounted on the roof supplies water to the upper stories. The tank is filled to a certain level by water from the main that is boosted by a pump in the basement. (*Note*: Water can be pushed up from below to just about any height. However, it can never be sucked up from above to any height greater than 33 feet, which is the static head equivalent of atmospheric pressure of 14.7 psi.) Since the roof tank supplies the upper floors, the pressure is determined by the height of the tank above a given floor, and not by the pump. Thus, the pressure at any level is constant. The disadvantage of the downfeed system is the considerable weight added to the roof, which results in a heavier and more expensive structure, particularly in earthquake-prone areas.

To avoid that, the *pneumatic tank* system uses a pressurized tank in the basement to supply higher levels. Some air is left in the tank, which can be compressed to act as a spring pushing up on the water in the tank (water does not compress). This takes up some floor space in the basement, and causes some air to be dissolved into the water.

The *tankless* system requires one or more variable speed pumps which constantly turn on or off or run at varying speeds to provide sufficient pressure at whatever demand rate the building requires. This tends to wear out the pumps more rapidly, but requires hardly any floor space and no additional roof space or structure.

Friction

The remaining consideration is the *flow rate*, and the resulting *pressure losses* due to *friction*. This is similar to the friction losses inside HVAC ducts as discussed in Lesson Three. The friction loss is a function of the diameter of the pipe and the flow rate through it. The smaller the diameter, the greater the friction at a constant flow rate. The greater the flow rate, the greater the friction at a given diameter. Valves, tanks, and other devices in the line also add friction. Thus, the vertical pipes in the building and the pipe from the street to the building represent additional losses (at design flow rate) which must be considered in addition to the lift requirements. The water meter itself represents a pressure loss depending on flow rate, from 1 to 25 psi for flow rates from 4 to 1,000 gpm.

Example #2

If there is an additional pressure drop of 12 psi due to friction in Example #1, what is the required water pressure at the base?

Solution:

P_{tot} = 52.2 psi (for lift)
     + 15 psi (for flush)
     + 12 psi (to overcome friction losses)
     = *79.2 psi (min)*

Hot Water Systems

There are usually two complete systems in a building, one which supplies cold water directly to the fixtures, while the other supplies cold water to a storage tank which is heated (the water heater), and then supplies the hot water to the faucets.

The water heater is always pressurized, and is rated in terms of its *volume* and its *recharge* rate. The volume is the capacity of the tank in gallons, and the recharge rate is the length of time that the tank will take to reheat itself after it has emptied out its volume of hot water.

There are several other variations. One is the solar hot water heater discussed in Lesson Two.

Another variation is the continuous loop system. Rather than allowing the water in the hot water pipes to cool off, thus requiring the user to waste all the water between the faucet and the heater before obtaining hot water, the hot water is continually pumped around a closed loop in the building.

There is a small, steady heat loss from the hot pipes, but no wasted water, since the user gets hot water almost instantaneously at the tap. In any case, it is important that all hot water pipes be insulated, since this is always the most cost effective first step, to conserve energy.

A totally different approach is the *in-flow* heater or *instantaneous* heater, in which only cold water is supplied to each fixture. The cold water is heated when the hot water faucet is turned on, or just prior to the expected need. This may be done automatically or manually. It is, of course, much more efficient, since no heat is wasted and no hot water is lost. The first cost is usually higher, and the convenience (flow rate, etc.) is often not as great, but such systems hold real promise. They may consist of electric resistance coils or small gas burners or heat exchangers.

Thermal Expansion

As with other building components, pipes expand and contract due to changing temperatures. This doesn't affect diameter very much, but it affects length a great deal. The change in the length of a pipe may be expressed by the formula:

$$\Delta L = Lk(T_2 - T_1)$$

where
   ΔL = the change in length
   L = length
   k = coefficient of expansion
   T_1 = original temperature
   T_2 = final temperature

TABLE 7.1—THERMAL EXPANSION COEFFICIENTS	
Material	**Coefficient**
Steel	6.5×10^{-6}
Cast Iron	5.6×10^{-6}
Copper	9.8×10^{-6}
PVC	35×10^{-6}

Example #3

The temperature of a 100 ft. length of copper pipe increases from 65°F to 160°F when hot water starts to run through it. How much does it expand?

Solution:

$\Delta L = Lk(T_2 - T_1)$
   $= 100 \text{ ft. } (9.8 \times 10^{-6}) (160° - 65°)\text{F}$
   $= 0.0931 \text{ feet}$
   $= 0.0931 \times 12 = 1.12 \text{ inches}$

That is quite a bit of movement and may present problems, especially if the pipe is fastened to a wall.

Therefore, pipe supports on long runs should be flexible, especially for hot water pipes. Some pipe supports actually have rollers, so that the pipe can move horizontally while it is supported vertically. Pipes should be supported every 4 feet for plastic, 6 feet for copper and 12 feet for steel.

VALVES AND FIXTURES

Valves

There are several kinds of valves in a plumbing system. A *gate valve* is intended to be entirely on or entirely off. It has a minimum restriction when fully open, but causes a great deal of turbulence when partially open.

Globe valves are used not only to turn water on and off but also to meter or throttle the flow at intermediate rates. Globe valves restrict flow even when wide open.

A *check valve* is essentially a *backflow preventer*. Its purpose is to prevent water from moving

GATE VALVE

GLOBE VALVE

ANGLE VALVE

Typical plumbing fixtures include lavatories, sinks, showers, and appliance hookups such as dishwashers and automatic icemakers. Such fixtures include a valve with a metering or flow restriction capability. These used to be similar to globe valves and were called *angle valves* or *screw* and *seat* or *washer* and *seat* valves. They screwed a washer down against a seat to shut the flow off, or opened and regulated the flow by screwing progressively away from the seat. These were twist handle faucets and hose bibbs. Now there is a myriad of patented single handle systems and cartridge systems which shut off more completely and easily when new, but which tend to cost a great deal when they need maintenance, and become automatically obsolete when the manufacturer "improves" the model. A new washer costs 70 cents, but a new cartridge costs $24.00, and a 15 year old cartridge is often impossible to replace.

backwards through the system. This is critical in avoiding contamination of the community supply, for example, to prevent backflow from an irrigation system. The simplest version consists of a flap which opens in the direction of the desired flow, but is pressed shut by flow in the other direction.

A preferable type is a ball which is spring loaded, so that it is pushed away from the mouth of the pipe by a desired flow, pops back with still water, and is pressed tightly shut by any reverse flow.

CHECK VALVES

AIR CHAMBER

SHOCK ARRESTOR

SURGE ARRESTORS

Pressure release valves are safety devices which keep systems from exploding by opening when the pressure exceeds some preset maximum. They are usually placed over a drain or in some area where the released water or steam will do no damage. They are required on water heaters, for example.

Surge Arrestors

One common source of annoyance is *water hammer*, that thumping or rattling sound which occurs when a faucet is shut off rapidly. This occurs upstream of the fixture where a long straight pipe run ends in a corner. The water which was flowing in the pipe is suddenly halted, and the inertia hammers the corner at the end of the run. In extreme cases it can actually burst out the end of the pipe. The solution is to insert a cushion or device with a damping effect somewhere in the system. These devices are called *shock arrestors* or *surge arrestors*. Alternatively, a length of vertical pipe with

nothing but air in it may be included in the system. The air compresses and bounces the water back at the original pressure, absorbing the shock.

Fixtures and Flow Rates

Different fixtures have different flow requirements. In order to determine the required size of pipe, an arbitrary unit is used for pipe sizing which takes into account the likelihood that all the fixtures will be in use at the same time. This is called the *fixture unit (FU)*. The relationship between gallons per minute (gpm) and FU is not constant, but varies with the number of fixture units. For example, 1,000 FU is equivalent to 220 gpm, yet 2,000 FU is not double that, but is only 1.5 times as much, or 330 gpm.

Two tables are necessary to determine pipe sizes: FU's per fixture type, and pipe sizes for total FU's.

TABLE 7.2—FIXTURE UNITS PER FIXTURE TYPE

Fixture	FU Value	Minimum Trap Size, Inches
Bath tub	2	1 1/2
Drinking fountain	1/2	1 1/4
Lavatory	1	1 1/4
Shower	2	1 1/2
Kitchen sink	2	1 1/2

Reproduced by permission of the International Conference of Building Officials.

than the faucet nozzle. This avoids the waste water being sucked into the faucet if there is a loss of pressure in the supply line, which could occur, for example, when several faucets two stories below are turned on full, and the pressure temporarily drops or becomes negative on this level. Where siphoning is likely, a *vacuum breaker* must be installed, for example, over a flushometer valve which could back up if there were a loss in the supply line pressure.

WASTE SYSTEMS

Another important consideration is that waste water must not be allowed to contaminate the supply water. For example, faucets should never extend down into the sink where standing water might cover the tip of the faucet. There should always be a two inch air gap designed into the system to prevent siphoning, by making the overflow on the sink or tub two inches lower

While the major consideration with water supply is to keep it uncontaminated, the major consideration with drainage is the opposite, that is, to keep it from causing contamination. For this reason, and for the benefit of everyone downstream, *sanitary waste* from inside buildings is usually kept separate from *storm drainage*, which is the runoff of rain water from outside buildings.

TABLE 7.3—PIPE SIZES FOR FIXTURE UNIT TOTALS

Pipe Size Inches	Drain or Sewer 1/8" Fall	1/4" Fall	1/2" Fall	Horizontal Fixture Branch	Stack of 3 Branch Intervals or less in Height	Stacks Greater than 3 Branch Intervals* Total for Stack	Total for One Branch Interval
1 1/4		1	1	0	0	0	10
1 1/2		3	3	3	4	8	2
2		21	26	6	10	24	6
1 1/2		24	31	12	20	42	9
3	36	42	50	20	48	72	20
4	180	216	250	160	240	500	90
5	390	480	575	360	540	1,100	200

*Branch Interval—A distance along a soil or waste stack corresponding in general to a story height, but not less than 8 feet, within which the horizontal branches from one floor or story of a structure are connected to the stack.

Reproduced by permission of the International Conference of Building Officials.

Sanitary Waste

Sanitary waste water is always assumed to be contaminated, because sometimes it is. This is because the organic matter in sanitary waste is in various stages of decay. This decay produces by-products which are dangerous to health, obnoxious in odor, and sometimes flammable, such as methane gas, and laden with bacteria which are likely to cause disease.

The trap under a sink, for example, is not just to catch your contact lenses before they disappear forever. The trap remains full of water even after the remainder of the waste has been transported away by the water flow. The water in the trap prevents the methane or sewer gas created by the decomposition of the sewage from passing back up the drain pipe into the occupied spaces in the buildings.

There are two categories of sanitary lines: *soil lines*, which carry the water from toilets, urinals, and similar fixtures, and *waste lines,* which carry all other waste water from inside the building.

To relieve the pressure that builds up, and also to break the suction or siphoning action that occurs when water passes down through the system, sanitary drainage is connected to vents which rise out of the building to the open air above. There are three types of venting: vent stacks, stack vents, and soil stacks. The *soil stack* is a large pipe into which all of the soil and waste lines from one or more levels empty. It is open to the outside air at the top. The *vent stack* is a smaller pipe which is the air intake line for all the fixtures, and which is also, separately, open to the outside air at the top. In the soil stack, the section *above* the highest fixture is called the *stack vent* (as opposed to vent stack) and vents the soil stack. The vent stack is a "stack" of vents, the stack vent is something that vents the top of the soil stack. If this is

confusing, we suggest that you look at the illustration on the following page. The minimum diameter for a vent is 1 1/4" or half the diameter of the drain it services, whichever is larger.

Typical Layouts

The material used most often for sanitary lines is cast iron pipe, although ceramic pipe is sometimes used for lines outside the building, and copper or galvanized steel may be used for vents. Plastic pipe may also be used, but is usually limited to residential applications.

The average toilet uses from three to five gallons of water per flush. Needless to say, that is quite a bit of clean water. This is true of *reservoir* or *tank type* toilets, which drain a tank through the waste bowl and then refill the tank, and of *flush valve* or *flushometer* toilets, which turn on water at a high rate of speed for a short time. In order to conserve water, several alternatives have been proposed, especially for residential applications where low volume is expected. The simplest alternative is the use of a smaller reservoir, which drops the water from a greater height, giving it a greater velocity.

More innovative is the *composting toilet*, in which no water at all is used, and the waste is stored below and vented to avoid noxious or unhealthy odors from collecting in the building. The biodegradable garbage from the kitchen also drops directly into the tank and eventually the combination provides a rich natural fertilizer. One example is the Clivus Multrum, a brand name for a Swedish company which manufactures composting toilets, which has a recycling time of approximately two years. After the toilet has been in use long enough, there is a steady supply of fertilizer.

Another, more complicated, concept separates urinal and toilet lines (soil lines) from sink and shower lines (called *grey water*). This does not

OPEN TO OUTDOORS

STACK VENT

VENT STACK

LOOP VENTING
(BRANCH VENT RETURNED TO STACK VENT INSTEAD OF VENT STACK)

ABOVE THE HIGHEST BRANCH DRAIN, A SOIL OR WASTE STACK IS KNOWN AS A STACK VENT.

CLEANOUT

SOIL LINE ✳✳

CONTINUOUS VENTING

TRAP

CLEANOUT

SOIL STACK ✳

CIRCUIT VENTING

FRESH AIR INLET

CLEANOUT

CLEANOUT

CLEANOUT

SLOPE

HOUSE DRAIN

CLEANOUT

HOUSE SEWER   3'-0"

✳ CALLED WASTE STACK IF THERE IS NO TOILET CONNECTED.

✳✳ CALLED WASTE LINE IF THERE IS NO TOILET CONNECTED.

SANITARY STACKS

reduce the amount of water per flush, but by keeping the organic wastes separate from the wash water, it allows the wash water to be processed and recycled on site, or at a small community scale. Although certain detergents must be avoided, this concept can be ecologically beneficial since it reduces the amount of potable water which is wasted.

HANDICAPPED ACCESS

Introduction

Although the standards for barrier-free architecture vary somewhat from state to state, the exam is based on the national standard, which is ANSI A117.1. Candidates should therefore become familiar with the handicapped provisions contained in that publication.

Rest Rooms

Toilet stalls for the handicapped in public rest rooms present a number of problems. To enable a person in a wheelchair to leave the stall, the door must swing outward, and he/she has to be able to open it. This usually means turning around, which requires a clear circular space of 5'-0" in diameter at a height of 10" above the floor. This is ideal and is in excess of the minimum ANSI requirements. Although stalls narrower than 5'-0" are permitted by the ANSI Standard, they present problems to a handicapped individual, who must be able to open a door behind his/her back.

In addition, once inside the stall, the individual must be able to transfer to the seat. It is best if the seat is at the wheelchair height of 1'-7", so that the return transfer is not to be "uphill." In stalls that are less than 5'-0", there must be grab bars on each side wall for the individual to lift himself or herself and turn around. Again, the 5'-0" stall is preferable, because it allows a side, or parallel transfer, where the wheelchair-bound individual backs up next to the seat and shifts across onto it. This also requires grab bars, one on the side wall and one on the rear wall. In both cases the grab bars should be set at a height of 2'-9" to 3'-0".

Lavatories

There should be at least one lavatory which has sufficient clearance underneath so that a wheelchair can be moved up under it. The fixtures should have large lever controls rather than a small handle, or most absurd, a single, little round ball. Exposed hot water lines under such sinks should be well insulated. The mirror above this sink should be tilted slightly forward, so that a seated individual can easily see himself or herself. It is also a good idea to place the faucets on the side of the lavatory instead of at the back, if the counter is deep.

Drinking Fountains

Two different drinking fountains are often used, one at a height of 36" to 39" for adults without physical impairment, and the other at 32" (preferred) to 36" (maximum) with a clear space below for wheelchair access. The lower fountain should protrude from the wall as far as possible, consistent with safe traffic flow. Some paraplegics dehydrate rapidly, and access to a drinking fountain is therefore important.

Baths and Showers

Handicapped people must be able to bathe or shower in hotel rooms, and there are many hotel rooms in which this is simply not possible. Bath tubs should be supplied with grab bars, and at least one tub in every hotel should be available with a seat, roughly at wheelchair height, or at least at the height of the tub edge. Elevating the entire tub slightly to achieve this is an excellent idea. At least one shower should have a minimum height curb, or no curb at all, and a door wide enough (33") to allow a wheelchair to roll in. Again, a seat at wheelchair height is preferable, as well as a flexible hose and nozzle arrangement for the shower head. If space permits, a shower stall with a clear space of 5'-0" in diameter is best. If necessary, the seat may intrude to create a 4'-0" circle, as long as 5'-0" is available at a 10" height above floor level, to clear the feet and the front wheels of the chair. Again, these dimensions are ideal and in excess of the minimum ANSI requirements.

In all of the above situations, common sense and a bit of thought are the primary considerations. Simply consider how a wheelchair-bound individual would go about using the facilities, and design accordingly.

MAINTENANCE

Unfortunately, every sanitary system eventually clogs up. However, there are ways of delaying that event, and there are fittings made to deal with the problem when it does occur.

Interceptors

Interceptors are designed to catch grease, hair, oil, string, rags, money, toothbrushes, etc., which get into the system. They are required by code for certain establishments such as restaurants, which produce sufficient grease to create problems for the sewage system and treatment plant. Every interceptor is provided with some means of cleanout access, so that trapped material may

be removed and discarded. Please note that a "trap" is the U-shaped line which keeps methane from the sewer system out of the building, while an "interceptor" is a device which keeps grease and other waste matter from the building out of the sewer system.

In addition to the cleanout for each interceptor, there are *cleanouts* for the system itself. A cleanout is just a Y-shaped segment of pipe which serves an area otherwise difficult to access, where one arm of the Y has a plug screwed in at its end. At least one cleanout is required where the building drain joins the sewer system. Cleanouts should be placed every 50 feet in pipes under 4 inches in diameter, and every 100 feet in larger pipes. There should also be a cleanout at every corner where the pipe changes direction more than 45°. When the system does clog up, the plug is unscrewed and a "snake," a length of flexible cable with a bit at the end, is inserted and rotated until the clog is broken up. The clog may be removed through the cleanout, or flushed down the sewer, or both.

GREASE, etc.

METAL, etc.

INTERCEPTOR

PLUG

OPTION OF INSERTING SNAKE

CLEANOUT

Manholes

Manholes are the equivalent of cleanouts for large lines, ten inches in diameter and greater. They occur at 150 foot intervals and wherever a new line is joined to an existing one. They also provide access for inspection.

SEWAGE TREATMENT SYSTEMS

Public Systems

In public sewage systems, all sewage is treated in a waste treatment plant before being returned to the nearest body of water. This is done by directing the flow into settling chambers which allow the solids to settle out. The remaining liquid sewage is treated using activated sludge, which is simply a rich mixture containing the necessary bacteria to digest the waste materials. The resulting clear water is chlorinated and returned to the surrounding water. The solid waste is put in an anaerobic digester (no oxygen present) and reduced in volume and digested by different bacteria. The result, *sludge*, is dried and either put in land fill, or occasionally used as fertilizer. Unfortunately, because of pollution today, sludge is usually contaminated, which precludes its use for most farming and residential applications.

When a public sewage system is unavailable, some form of private sewage treatment is necessary, such as a septic tank with leach field or a cesspool.

Cesspools

Cesspools provide the cheapest sewage treatment, but also the least desirable, and have been outlawed by code almost everywhere. A cesspool is simply an underground chamber with a porous bottom and porous walls. The sewage soaks into the surrounding ground until everything gets clogged up. This means that eventually a new cesspool must be dug and the sewage lines rerouted to it.

Septic Tanks and Leach Fields

A *septic tank* is a lined chamber, or often a steel tank, into which the sewage collects. The solid material deposits out, and the liquid waste passes on into a leach field. The solid material must be removed every few years. Septic tanks are sized based on an expected flow of 100 gallons per day per person, with a minimum capacity of 500 gallons.

The *leach field* or *tile drain field* is simply a grid of ceramic pipe laid underground, not quite together end to end, so that the liquid leaks out, over a bed of gravel, which filters the liquid waste before it seeps into the soil. Where the soil is impermeable, a large basin is dug and filled with sand. The liquid waste leaks into the sand, and after it is filtered by the sand, it is collected at the bottom of the basin, chlorinated, and returned to a local body of water.

STORM DRAINAGE

Surface runoff from rainfall is generally kept separate from sanitary waste because it is basically clean, and does not need to be treated in the same way as the more hazardous wastes. If it were piped into the sanitary system, it would overload the system. This sometimes happens with older systems or when rain water is illegally drained into the sanitary system. At such times, sewage treatment plants allow the raw sewage to flow along with the rain water out into the open stream or ocean discharge.

Water Table and Ground Water Recharge

Before urbanization covered much of the land with buildings, parking lots, and streets, most of the rain water soaked into the ground before running off. Thus, there was a steady flow of *ground water* to *recharge* the *water table* (underground water level). It also meant that the likelihood of a local flood was lower than today. The dual problems of insufficient water retention for wells and springs and excessive runoff and flooding can be ameliorated by intelligent handling of storm drainage at the building and site.

Many suburban corporate campuses now have large depressed recreational areas, which are allowed to flood during heavy rainfall. Much of the water in the flooded area soaks into the ground, while the flood crest passes. These areas must still be properly drained, to avoid breeding mosquitoes, but at a limited rate. This drainage is done by means of *swales* and *catch basins*.

Swales

Swales are shallow V-shaped sloping channels in the grass which take the surface runoff to points where it may be collected and/or disposed of.

Catch Basins

Catch basins are similar to manholes, except that the top has a grate instead of a cover. They are placed at the lowest point in a swale or a depression, to collect the runoff and pass it into the storm drainage system, which empties into the local stream or lake.

SUMMARY

In this lesson we have discussed water supply, sanitary waste and storm drainage. We have identified some typical materials and connections, typical problems, and common systems, fixtures and valves.

Although exam questions may sometimes require calculations, more often candidates are tested on their understanding of basic concepts and their ability to recognize the various elements of water systems.

LESSON 7 QUIZ

1. Select the correct statements.

   I. Rain water is slightly acidic.

   II. Rain water is slightly alkaline.

   III. Rain water tends to corrode metal pipes.

   IV. Rain water is often less healthful than ground water.

   A. I and III
   B. II only
   C. I, III, and IV
   D. II, III, and IV

2. Which of the following statements is correct?

   A. Ground water is naturally soft.
   B. Hard water is often hazardous to humans.
   C. Ground water is naturally hard.
   D. Hard water can be controlled by the use of acid neutralizers.

3. Select the correct statement.

   A. Acidic water causes deposition, so pipes should be oversized.
   B. Acidic water can be controlled by using the zeolite or ion exchange process.
   C. PVC pipe is appropriate for supply piping in exposed locations above ground.
   D. Copper and PVC are resistant to corrosion and therefore appropriate where water is acidic.

4. Which of the following represents an extremely acidic pH reading?

   A. pH = 5
   B. pH = 7
   C. pH = 9
   D. pH = 17

5. Which of the following piping materials is heated when two pieces are joined?

   A. Steel
   B. Copper
   C. Iron
   D. Plastic

6. If the street main provides 100 psi, and there is a loss of 10 psi due to friction between the street and the building, how high can we place a faucet if the required pressure at the fixture is 10 psi?

   A. 74 feet
   B. 98 feet
   C. 124 feet
   D. 184 feet

7. A downfeed system is different from a pneumatic system in that

   I. a downfeed system is necessary only when the street main pressure would not supply the upper floors.

   II. a downfeed system is not pressurized.

   III. a downfeed system has no storage tank.

   A. I only
   B. II only
   C. III only
   D. I and III

8. Why is it necessary to keep storm drainage separate from sanitary sewage?

   I. Storm drainage is more polluted than sanitary sewage.

   II. Storm drainage tends to flow at high volumes, which are not treatable by the sewage plant.

   III. Storm drainage is not pressurized but sanitary drainage is.

   A. I only
   B. II only
   C. III only
   D. II and III

9. The purpose of a sanitary trap is to

    A. catch valuable objects before they enter the sewage system.

    B. catch dangerous chemicals before they enter the sewer system.

    C. keep storm drainage and sanitary sewage separate.

    D. keep sewer gas from backing up into the building.

10. What is a catch basin?

    A. An inlet into the drainage system for surface runoff

    B. A device which keeps grease and other waste matter out of the sewer system

    C. A type of cleanout which provides access for inspection and maintenance

    D. A lined chamber into which sewage collects

FIRE SAFETY

INTRODUCTION

Fire protection in buildings is the subject of this lesson, and two different codes are the source of most of the information. The Uniform

Building Code (UBC) is a model building code which specifies the fire protection which must be incorporated into any building. The features which are covered include fire ratings of walls, floors, and roofs; and the requirements for sprinklers, exits, standpipes, etc.

The sixteen volumes of the National Fire Code specify details of fire protection systems, particularly Volumes 10 through 13. In fact, they go beyond the purview of the architect in detailing how a building is to be used and how equipment is to be maintained.

While there are other fire protection codes, this lesson is based on these codes, since they are referred to by the NCARB. The candidate should have access to a copy of the current UBC as he or she studies this lesson.

FIRE SAFETY PRIORITIES

Traditionally, fire protection codes have had three goals. The first is to afford protection or escape for the occupants of a building. This may be accomplished by either of two means: evacuation from the building (egress), or moving the occupants from an endangered area to a protected area elsewhere in the building, a *place of refuge*. In small structures, evacuation is the usual approach, but in high-rise structures a place of refuge is the only practical method.

The second aim of fire protection is to insure sufficient structural integrity in the building so that fire fighters may enter and fight the fire without excessive risk of being trapped or injured by collapsing portions of the building.

The third and least significant fire protection goal is to allow the building to survive a fire so that it may be economically restored afterwards.

A fourth goal which is becoming a major thrust of building codes is to prevent fires from starting, or once they have started, to extinguish them immediately and automatically. In the MGM Grand Hotel fire in Las Vegas, parts of the casino and hotel were completely destroyed, and there was significant loss of life. Yet immediately adjacent to the point of origin of the fire, a restaurant which was sprinklered survived the fire with tablecloths intact.

BUILDING CODE

In order to determine the UBC requirements for a specific structure, the architect must classify the building by its function, its construction type, and its location.

Occupancy

To determine the degree of fire resistance required in a given structure, the UBC has established a number of specific occupancy classifications. These are listed in Table 3-A of the UBC, which the candidate should review. Generally, however, these occupancies may be categorized as follows:

For basic classification, all buildings are assigned a *group* letter, and then for more specific definition, a number which identifies the subcategory or *division* within the basic

TABLE 8.1—OCCUPANCIES	
Group	**Description of Occupancy**
A	Assembly
B	Business
E	Educational
F	Factory and Industrial
H	Hazardous
I	Institutional
M	Mercantile
R	Residential
S	Storage
U	Utility

classification. Open parking garages, for example, are Group S Division 4, or S-4 occupancy.

Construction Type

Buildings are also categorized by their type of construction which, in turn, determines their degree of fire resistance. The candidate should be aware of the requirements for the major building elements for each type of construction, which are specified in Table 6-A of the UBC.

The basic method of assigning fire resistance ratings is to test each construction material or assembly under standardized conditions for a period of time, typically one to four hours. The detailed requirements for each type of construction are specified in Chapter 6 of the UBC, but the candidate need not study this chapter in depth, skimming should be sufficient. The basic idea is that there are five construction types, ranging from the most fire-resistive *(Type I)*, to conventional wood stud construction *(Type V)*.

The location of a building on its property with regard to building setbacks, alleys, public streets, and property lines also affects the required fire resistance ratings of exterior walls. In this regard, the candidate should review UBC

Table 5-A. The basic intent of the requirements is that fire should not be allowed to spread from one building to the next, so increased fire resistance is required at property lines, etc. This gives the local fire department time to put out the fire in a single building before it consumes an entire block of buildings. For the same reason, openings in exterior walls are limited in size, so that fire cannot easily pass through the rated walls.

The maximum floor area of a building is limited in accordance with Table 5-B of the UBC. Buildings of more fire-resistive construction are allowed greater floor area. In fact, buildings of Type I construction are permitted unlimited floor area for most occupancy groups. It is also important to note that the installation of an automatic sprinkler system in an occupancy which does not otherwise require it permits a very significant increase in allowable floor area.

Similarly, the height and number of stories in a building are limited for various types of construction, depending on whether or not the building is sprinklered.

The number of occupants assumed to be in a building is based upon the building's occupancy and floor area.

To summarize, a building must comply with the detailed requirements of the code which specify fire-resistive ratings for walls and other elements, height, and exits for the different occupancies and types of construction.

COMPARTMENTATION

Frequently, more than one occupancy group is located within a given structure. In order to permit such mixed occupancies, the construction of the walls separating them is also specified. For example, if a retail store (M occupancy) is in the same building as a repair garage (S-3 or H-4 occupancy), a one-hour separation is required between the two, in accordance with Table 3-B of the UBC.

Similarly, when a design requires more floor area than is permitted for its occupancy, the space may be separated into two or more portions, each of which must comply with exiting and other requirements as if it were a separate building. The walls and floors separating the compartments must then have a fire rating of four hours or two hours, depending on the building type.

EACH COMPARTMENT IS ENCLOSED BY FIRE RATED WALLS, FLOORS, DOORS, ETC.

FIRE

PLACE OF REFUGE

COMPARTMENTATION

In order for the compartmentation to be effective for fire safety, all openings through the walls must be closed with fire rated devices.

This includes not only doorways and windows, but also air ducts which pass through the fire separations. In order to have a fire rating, a prototype of the entire fire assembly must have been previously tested and approved by Underwriters' Laboratories (UL) or a similar, recognized testing agency. This test is done by burning a specified fire on one side of the assembly for the specified time and then testing its function. With doors, for example, this means shooting a stream of water from a fire hose at it to see if it pops out of the frame. Note that not just the damper or the door must be fire rated, but the entire assembly—including the frame, the mounting hardware, the latches, the hinges, the sill, and the door itself. In addition, fire assemblies must be either self-closing or automatic if subject to an increase in temperature or to products of combustion. Self-closing assemblies are typically held open by means of fusible links which melt when the ambient air temperature exceeds a certain value, often 165°F. The assembly will then close and latch by means of a spring or gravity device.

EXITS

The code requires that exit passageways be provided in every building from every part of every floor to a public street or alley. This means having at least one exit from every building or compartment of a building, and when the rated number of occupants exceeds a certain number, two or more exits are required. For example, assume that a dance studio occupies a 40 × 50 foot space on the second floor of a store building which has separated stairways to the street. Its occupancy is A-3, and it is assigned a maximum capacity of (40 × 50)/7 = 286 persons (see Table 10-A). Since any dance floor with more than 50 persons requires two exits, this space requires two legal exits. Most buildings require *two or more exits*. The reasoning is that if there were

only one exit, it could be blocked by the fire and the occupants would be trapped. If there are two means of egress, it is unlikely that both would blocked by fire.

Each exit stairway must be within 150 feet of any point (200 feet in a sprinklered building). These distances are increased to 200 feet and 250 feet respectively in the 1997 UBC. Thus, a 300-foot-long building requires at least three exit stairways (one in the middle and one at each end).

Since an exit passage may be the escape route for several spaces, the total flow in the passage must be considered. This is done by determining the total occupant load to be served by the passage and multiplying by 0.2. The result is the minimum exit width in inches. Under no circumstances should any part of an exit passage be less than *44 inches wide*.

All of the doors in the exit passage must swing in the direction of travel, and must be unlatched or activated by *panic hardware*. Panic hardware is a door-latching assembly which will release the latch if a force not exceeding 15 pounds is applied to it, such as would occur if someone stumbles blindly into it in the direction of travel. The assumption is that the hallway may be filled with smoke, and the occupants may not be able to see (or think) properly.

Exits from all public spaces must be provided for handicapped persons. This may mean ramps or paths to safe compartments. Elevators may only be used if they are not connected to heat responsive call buttons, for obvious reasons. In general, elevator shafts collect smoke, somewhat like a chimney, from the floor that is on fire, and their use should be avoided if at all possible.

All exit stairways must be of fire-resistive construction. Stair enclosure walls in buildings which are four stories and greater in height, or of Types I and II-F.R. construction, require a minimum of two-hour construction. One-hour rated walls are acceptable for all other buildings. In buildings greater than 75 feet in height, the entire required exit enclosure must be pressurized, with pressurization occurring automatically upon activation of the fire alarm system.

Large floors should be subdivided so that the first means of escape is simply to get across the division to the other section of the same floor. This limits the spread of fire, so that fire fighting personnel and equipment can more easily control the fire, and is a more satisfactory means of egress for handicapped individuals than moving to another floor. Again, all doors and dampers in the division boundary wall must be fire-rated.

CLASSES OF FIRE

Fires have been divided into four classes: A, B, C, and D. Class A fires are those involving ordinary materials, including wood, paper, cloth, and rubber. These can generally be extinguished with water, or with an extinguisher which sprays water. Class B fires involve flammable gases and liquids, such as natural gas, gasoline, oil, etc. These tend to float on top of water, making water ineffective in extinguishing them. Class C fires involve electrical equipment, and the extinguishing medium must be electrically non-conductive, and therefore water is not acceptable. After the source of electricity is disconnected, Class A or B extinguishers may be used. Class D fires involve combustible metals such as sodium, potassium, magnesium, etc., and require special extinguishers. In fact, sodium at room temperature may actually burst into flames on contact with water.

Special Extinguishing Media

There are several special media used in hand-held and automatic fire extinguishing systems. One such medium is Halon (Halon 1301 or Halon 1211), which is not toxic for brief exposures, and which may be used safely on Class B and C fires. However, prolonged exposures should be avoided. It displaces oxygen, which is useful in fighting fires, but it can eventually result in asphyxiation. Carbon dioxide (CO_2) also displaces oxygen. When either of these is used in an automatic system, locally audible and visible alarms must be provided which warn personnel to leave the vicinity. Both of these extinguishing media are preferred in areas where there are valuable documents or artwork. Halon is particularly common in computer installations. Neither equipment nor records are damaged, but the fire is smothered. The cause of the fire should be found immediately, however, as it is likely to restart when oxygen returns.

FIRE DETECTION

There are three forms of fire detection available. They are based on ionization, photoelectric detection, or temperature sensing.

Ion Detectors

The *ionization detector* responds to the chemical *products of combustion (POC)* present in the air during a fire, even in the earliest stages. These may be visible or invisible; the ionization detector is sensitive to both. Such detectors are now inexpensive and available even for home use. Batteries should be checked periodically. The one problem with ion detectors is that they also detect smoke from the kitchen or cigarette smoker.

Photoelectric Sensors

The *photoelectric detector* reacts to visible smoke in the air which blocks a beam of light. This may measure across a larger volume of air, but may also miss some of the early signs that the ion detectors will pick up. Given the low cost of ion detectors, they have surpassed the photoelectric detectors in popularity.

Both systems are actually preferable to waiting for human detection of fire. They are capable of sensing smoldering fires long before the fire is visible to the naked eye, which tends to be too late with many kinds of fire.

Small or smoldering fires are particularly dangerous because of a phenomenon known as *flashover*. Smoldering fires release gases which are at fairly high temperatures, and which collect near the ceiling. The ceiling materials become extremely hot over a rather broad area. When they finally reach combustion temperatures, they tend to do so all at once, and a very small fire can become a huge one in moments. In extreme cases, the gases then superheat, almost exploding out of the area where the fire started.

All detectors should be placed close to or on the ceiling. Vertical circulation spaces in the building often provide good locations. For example, smoke created in one portion of a home will often rise up a stairway.

Heat Actuated Sensors

A less sensitive detector is the heat actuated type. There are several ways this is done, from the original simple fusible link to more sophisticated electronic devices. At the simplest, a piece of wax or paraffin separates the contacts of an alarm circuit, and when it becomes warm and melts, the alarm goes off. Similarly, fire separation doors in older buildings were spring loaded to shut, or to drop, and the only thing holding them open was a piece of wax. When the wax melted, the doors shut. Although primitive, this is still surprisingly effective, and has saved numerous buildings, especially some of our historical landmarks.

Such devices rarely cause false alarms, but often are actuated too late to save the room in which the fire began.

Several other functions may be set in motion by the fire alarm. These may include:

1. Actuating remote alarms, in the building or even at the local fire station.
2. Actuating other extinguishing systems.
3. Overriding elevator controls.
4. Closing fire doors, fire dampers, and otherwise controlling smoke and fire migration.
5. Varying fan speeds throughout the mechanical equipment system.

Modern buildings are often equipped with fire fighting stations, which become active in case of a major fire. Elevators and intercom systems may be controlled from the station. Some are provided with information from the fire detection devices as well, so that the fire fighters may know the progress of the fire throughout the building at a glance.

STANDPIPES

The normal water distribution system in a building is rarely adequate to fight a fire within the building. For this reason, *standpipes* are used. These are not intended to go to each space, but rather to distribute large volumes of water to each floor, from which the fire fighting hoses and equipment can distribute the water to the spaces where it is needed. There are two types of standpipes, dry and wet.

Dry Standpipes

Dry standpipes are large diameter water risers which are normally empty and not connected to a water supply. The lower end terminates at street level where the fire department can conveniently connect it to a fireplug via a

DRY STANDPIPE

pumper truck, which is capable of pumping the water up through the pipe.

The fitting at the lower end of the standpipe is referred to as a *siamese fitting* and is arranged to accept either two- or four-hose connections from fire department pumpers, depending on whether the dry standpipe is four or six inches in diameter. A 2 1/2 inch outlet connection must be provided at every floor level higher than the first floor and at the roof. If the dry standpipe extends more than 75 feet above grade, the pipe connections must be provided in every required stairway as well.

The dry standpipe is, in effect, a portion of the fire department's equipment, which has been permanently installed in the building. To use it, the fire fighters connect a pumper to the siamese fitting at the lower end, and then carry their own rolls of hose up inside the building, connecting them to the 2 1/2 inch outlets wherever necessary.

One of the benefits of the dry standpipe is that there is no need to worry about rusting or freezing. To insure that it remains dry, an automatic drain valve (called a *ball drip*) is located at the lowest point of the system.

Wet Standpipes

Wet standpipes are required in buildings of four or more stories in height, throughout most theaters and other places of assembly, throughout all hazardous occupancies, and throughout all Group I, B, S, and M occupancies. Wet standpipes are provided primarily for the use of the occupants of the building, but they must also be equipped with siamese fittings so that the fire department may supply additional pressure and flow. Wet standpipes must be located in a building so that every point of every floor is within 30 feet of the end of a 100-foot hose attached to an outlet. These hoses are usually pre-attached and stored folded in a wall case with a glass panel making them clearly visible.

ROOF

WET STANDPIPE RISER WITH 1½" OUTLET AND 100' HOSE AT EACH FLOOR

SIAMESE FITTING

MAIN BASEMENT METER

WET STANDPIPE

The wet standpipe system must be designed to supply at least 35 gpm at 25 psi minimum for at least 30 minutes. The water supply system itself must be adequate to provide 70 gpm for 30 minutes at 25 psi minimum. The water supply may be a pressure tank, a gravity tank, or an automatic pump as long as the power source for the pump is safe.

Buildings which exceed 150 feet in height require a *combination standpipe* for every stairway which extends from ground to roof, which replaces the pipe otherwise required. The combination standpipe must be equipped with the $2^1/_2$ inch outlets for the fire department and also with $1^1/_2$ inch hose racks as in a wet standpipe system.

SPRINKLER SYSTEMS

Automatic sprinklers are a widely-used and very effective means of extinguishing, or at least controlling, fires in their early stage. Being automatic, they do not require the attendance of a person for operation, thus providing protection even when a structure is not occupied. The actual flow of water from a sprinkler head is comparable to a very heavy rain storm, and therefore is not sufficient to cause injury to human occupants. However, sprinkler systems can damage a building's contents, such as paper, records, electronic equipment, artwork, or merchandise. Flow rates range from 5 to 20 inches of water per hour, depending upon code requirements.

The UBC requires automatic sprinklers in many occupancies and encourages them in others. With some exceptions, sprinklers are required in all of the following occupancies: basements and cellars of all buildings, except private houses and garages; the backstage area, dressing rooms, workshops, and storage areas of theaters; and any concealed spaces above stairways in schools, hospitals, institutions such as prisons, and places of assembly such as theaters and arenas. Automatic sprinklers are also required over all rubbish and linen chutes, except in private dwellings; in retail sales areas over 12,000 square feet per floor or over 24,000 square feet gross; and in all places of assembly over 12,000 square feet in area. The code permits an increase in allowable floor area and height, and allows wider spacing of exit stairs for sprinklered buildings. In many jurisdictions, the requirements for automatic sprinklers are more stringent than those in the UBC.

Wet and Dry Systems

The simplest type of sprinkler system consists of a pattern of sprinkler heads, each of which is equipped with a fusible plug or fusible link. In the event of a fire or very high ambient temperature, the fusible plug will melt and the water pressure in the pipe will cause a spray of water through the sprinkler head. Some heads are inset into the ceiling to be less conspicuous. Such heads often also pop out when activated to give better distribution. As with standpipes, there are both wet and dry systems.

WET PIPE SPRINKLER SYSTEM

The wet pipe system has the advantages of quick response and low initial cost. However,

DRY PIPE SPRINKLER SYSTEM

its disadvantages include the possibility of freezing and unnecessary wetting of building contents. To overcome the freezing problem, the dry pipe system was developed. In this system, the sprinkler piping between the dry pipe valve and all of the sprinkler heads is empty of water and is filled with compressed air. The pipe valve may be located in a warm enclosure. The disadvantage is that when the sprinkler is first activated, nothing but compressed air comes through it until the system between the valve and the sprinkler has been flushed of air. This may cause a dangerous time delay in long pipe runs.

Preaction System

The preaction system is a variation of the dry system which requires that both the sprinkler head be activated and an independent fire sensing device triggered. This avoids accidental discharge of the sprinklers and resultant water damage. It is not as fail-safe as either the wet or dry system, however. All of the above systems cause only the activated sprinkler heads to release water.

PREACTION SPRINKLER SYSTEM

Deluge System

The *deluge system* is based on the idea that if there is a fire somewhere within the space, wetting the entire space is the safest course of action. This is generally reserved for areas of high fire hazard. In the deluge system all of the sprinkler heads are wide open at all times, but the pipe system is empty of water. Release of the water is actuated by a heat or fire detection system installed in the area to be protected, which activates a valve, flooding the system with water.

DELUGE SPRINKLER SYSTEM

All sprinkler systems must also have a *siamese connection* outside the building to permit the fire department to augment the overall flow.

Hazard Levels

In order to determine how great an area may be served by one sprinkler head, three major hazard levels have been established. *Light hazard* identifies areas where the quantity of combustible materials is relatively low. Typical light hazard occupancies include churches, hospitals, museums, offices, and residential occupancies. A greater degree of hazard is identified as *ordinary hazard*, which is subdivided into groups 1, 2, and 3, with progressively greater hazard as the group number increases. Group 1, for example, includes automobile garages and laundries. Group 2 includes large stack room areas of libraries and printing and publishing plants. Group 3 includes paper processing plants and tire manufacturing plants. *Extra hazard* is the greatest degree of hazard and applies to areas such as aircraft hangars and explosives handling areas.

Sprinkler heads *must never be repainted*. This ruins the temperature sensing capability of the fusible link, and may cause the mechanism to be jammed by the paint.

Insurance Company Requirements

Because insurance companies base their rates, in part, on the adequacy of the sprinkler system in an insured structure, they always require, as a condition of the insurance contract, that the owner immediately notify them of any proposed or actual changes in the sprinkler system protecting the structure. Failure to do so may result in loss of coverage.

SUMMARY

We have discussed the basic principles of fire safety in buildings. There are several strategies, which fall into two groups: (1) fire-resistive construction and safe egress, and (2) preplaced fire-fighting equipment. Fire-resistive construction involves occupancy groups, construction types, fire-resistive ratings of floors and walls, exiting requirements, and maximum allowable floor areas and heights. Preplaced fire-fighting equipment includes fire detection equipment, extinguishers, standpipes, and sprinkler systems. As always, understanding the principles is better than memorizing tables.

LESSON 8 QUIZ

1. Compartmentation refers to
   A. sealing special chemicals in compartments, which are released when fire is detected.
   B. containing the fire in the area in which it began.
   C. dividing a sprinkler system into compartments based on required flow rate.
   D. dividing the city into Fire Zones 1, 2, and 3.

2. The minimum exit corridor width is
   A. 36 inches.   C. 44 inches.
   B. 39 inches.   D. 48 inches.

3. The maximum distance from a room to a fire stair exit is
   A. 75 feet.   C. 150 feet.
   B. 130 feet.   D. 200 feet.

4. If a wet standpipe system is installed, every point on every floor must be within how many feet of a connection?
   A. 75 feet   C. 150 feet
   B. 130 feet   D. 200 feet

5. Flashover occurs when
   A. all the materials reach combustion temperature at once.
   B. one building's roof ignites the next building's roof.
   C. a fire thought to be extinguished restarts itself from the embers.
   D. high temperatures trigger a sprinkler head.

6. Building occupancy groups are determined by
   A. the construction materials used in the structure.
   B. the number of people expected to use the building.
   C. the function for which the building is designed.
   D. the size of the building.

7. Dry standpipes are
   A. used by building occupants to put out a fire before it gets out of control.
   B. filled with Halon gas or carbon dioxide, and triggered automatically.
   C. used by fire fighters, who have to pump water into them.
   D. used to drain floors of excess water during fire fighting.

8. The major benefit of using sprinkler systems is that
   A. they do not require that someone be in attendance to be effective.
   B. they are considerably less expensive than standpipe systems.
   C. they do not damage valuable artwork, books or records.
   D. they may also be used as domestic plumbing systems.

9. Which of the following fire detection devices will sense the products of combustion, whether or not the fire is very large?

A. Ionization detector

B. Smoke detector

C. Sprinkler head with fusible link

D. Heat detector

10. Which of the following are required to have a siamese connection at the base?

I.   Wet standpipes

II. Dry standpipes

III. Wet sprinkler systems

A. I and II

B. II and III

C. I and III

D. I, II, and III

GLOSSARY

The following glossary defines a number of mechanical and electrical terms, many of which have appeared on past exams. While this list is by no means complete, it comprises much of the terminology with which candidates should be familiar. You are therefore encouraged to review these definitions as part of your preparation for the exam.

A

Absorption Coefficient   The percentage of incoming energy which is absorbed. In measuring radiant energy (light or heat) it is a unitless ratio which may vary depending on wavelength. In acoustics, it is called a sabin and may vary depending on frequency.

Absorption Refrigeration   A cooling process resulting from the absorption of vapor by a brine solution that is then heated to remove the moisture. The heat may be supplied by solar or other heating sources.

AC   The abbreviation for either air conditioning or alternating current, depending on the context.

Acoustic Power Level   See Power Level.

Air Gap   An unobstructed vertical path, open to the atmosphere, separating the outlet of a faucet from the overflow rim of the fixture it serves. The purpose is to prevent a momentary vacuum in the supply pipe from siphoning water from the fixture back into the supply pipe.

Alternating Current   An electric current that reverses its direction at regular intervals, generally 60 times per second in the United States. A plot of the voltage over time is a sine wave.

Ambient   Relating to a general or all surrounding condition. In thermal processes, it refers to the air temperature, as distinct from that of surfaces or objects. In lighting, it refers to the background light level, and in acoustics, it refers to the background noise level.

ASHRAE   The abbreviation for the American Society of Heating, Refrigerating, and Air Conditioning Engineers, the source of most of the standardized information on the subject.

Aspect Ratio   The ratio of the longer to shorter dimension of an air conditioning duct, that affects duct friction, or of a room, that relates to light reflection.

B

Ball Drip   The automatic drain valve at the base of a dry standpipe.

Blow Down   The drain that removes dirt that builds up in the floor pool of an evaporative cooling tower.

BTU   The abbreviation for British Thermal Unit, a unit of heat energy, which is defined as the amount of heat required to raise the temperature of one pound of water by one degree Fahrenheit.

BTUH   The abbreviation for Btu's per hour, an energy flow rate.

Building Automation   The control by automatic equipment of many functions in a large building, usually including the HVAC system, the fire detection and alarm system, and building security.

C

CFM   The abbreviation for cubic feet per minute, which is the flow rate of air in a mechanical system or duct.

Chiller   A piece of equipment which cools water for use in an air conditioning system.

Chill Factor   A fictitious temperature assigned to a combination of actual temperature and wind velocity that has the same physiological effect as still air at the chill factor temperature.

Chlorination   The addition of small amounts of chlorine to a water source to kill bacteria.

Circuit Vent   In sanitary drainage, any vent that serves two or more traps.

Clerestory   A window above eye level that admits daylight, such as the high windows in a cathedral.

Code   An organized body of rules and regulations adopted and enforced by a governmental unit, such as a building code.

Coefficient of Performance   The ratio of the amount of heat energy delivered by a heat pump to the amount of energy supplied, or the ratio of the amount of heat energy removed by a refrigeration machine to the amount of energy expended in its removal. It is similar to efficiency, but often exceeds 100 percent, and is therefore expressed as a number, i.e., 2.7, which is typical for a heat pump.

Coefficient of Utilization   The ratio of useful light arriving at the work plane to the amount of light emitted by the source. The CU depends on the reflectivity of different surfaces and the aspect ratios of the ceiling, wall, and floor cavities.

Comfort Zone   The combination of thermal and environmental conditions within which a human is comfortable, often shown on a psychrometric chart.

Conductance (C)   The rate at which a specific thickness of a given material conducts heat.

Convection   The heat transfer process that occurs when a warm fluid rises, displacing cold fluid which then falls.

Cycle   One complete set of repeating events, typically used with alternating current or sound.

Cycles per Second (CPS)   A measure of frequency in electric current or acoustics, i.e., the number of cycles per second of a wave or oscillation. In acoustics, the term has been largely replaced by Hertz, where 1 cps = 1 Hz.

D

Daylighting   The use of natural light from outside to replace or augment electrical light outdoors, which produces energy savings.

dBA   A decibel measured in the A scale, that is weighted to account for the special sensitivities of the human ear.

Decibel (dB)   A logarithmic measure of sound intensity level.

Degree Day (DD)   The amount by which the average outdoor temperature at a particular location is below 65°F for one day. Degree days may also be summed and stated for a month or a year.

Delta Connection   A method of connecting windings on a three phase transformer, end to end, that results in a triangular shape.

Dew Point Temperature   The temperature of air at that the water contained in the air begins to condense and form dew. The dew point for a given air sample is always lower than or equal to its current temperature.

Diffuser   A device through which the air from a duct enters a room, or a device through which the light from a fixture enters a room.

Dry Bulb (DB)   The temperature of air as read on an ordinary glass thermometer.

Dry Pipe Sprinkler   A sprinkler system whose pipes are normally pressurized with only air, thus being invulnerable to freezing temperatures. Upon actuation, the air is vented and supply pressure forces water through the system.

Dry Standpipe   See Standpipe.

E

Economizer Cycle   An energy-saving strategy in which a part of the HVAC system is shut off while the rest is used, such as shutting off the refrigeration when the outside air temperature is low, while the fan continues to operate.

Effective Temperature   A fictitious temperature having the same physiological effect as air of a standardized temperature, humidity, and velocity.

Efficacy   The ratio of the lumens emitted by a lamp to the electrical power consumed by the lamp.

Emissivity   A factor that represents the rate at which a given surface material gives off or emits radiant energy. The emissivity varies from 0 to 1.0, where 1.0 is the theoretical emissivity of a perfect black box at the same temperature.

Enthalpy   The total of sensible plus latent heat stored in the air. It is also known as Total Heat.

Exit   A continuous and unobstructed means of egress to a public way. Its minimum width is generally 44 inches.

F

Fire Assembly   A complete fire-resistive assembly consisting of a fire door, fire damper, or fire window and its mounting frame and hardware. The entire assembly, not just its components, must be approved and labeled by a testing agency that inspects the materials and workmanship during fabrication at the factory. Available ratings are 3/4, 1, 1-1/2 and 3 hours.

Fire Door   See Fire Assembly.

Fixture Unit   A unit of liquid flow used in sizing both supply and drainage pipes.

Flame Spread Rating   A numerical classification indicating the rate at which flame will spread in or on a given material, in which higher numbers flame up more rapidly.

Flushometer Valve   A valve that releases a definite amount of water into a plumbing fixture each time it is actuated.

Footcandle   The basic unit of illumination arriving at a work plane. One footcandle is equal to one lumen per square foot.

Forced Air System   A heating or cooling system that uses a fan to circulate heated or cooled air through ducts to the occupied spaces.

Forced Convection   The movement of a fluid by a fan or a pump, in order to force heat exchange.

Four Pipe System   A hot and chilled water system having separate return lines for each supply line, and with no mixing of the two streams.

Frequency   The number of cycles that occur per second, either in alternating current or acoustics. In acoustics, the frequency determines the pitch.

Frost Line   In a given location, the maximum depth in soil that is expected to freeze in cold weather. Water piping must ordinarily be buried below the frost line to protect against freezing.

Fusible Link   A piece of wax or paraffin that melts at a predetermined temperature, setting off a sprinkler head, an alarm system, or otherwise actuating a fire protection device.

G

Globe Thermometer   A thermometer that measures Mean Radiant Temperature (MRT).

Ground   An electrical conductor connected to the earth or to a pipe extending into the ground, used to dissipate hazardous currents into the earth.

H

Halon   A gaseous fire extinguishing medium that smothers fires, often used in automatic systems in computer rooms.

Heat Pump   A refrigeration loop used to bring heat into a space instead of removing heat from it. The term is also used for an entire system of such units attached to a recirculating heat sink.

HEPA Filter   A high efficiency particulate air filter, which removes dust and other tiny particles from a moving air stream.

Hertz (HZ) The frequency of a sound, or of an alternating current, equal to the number of cycles per second.

High Intensity Discharge (HID) A family of lamps consisting of a quartz envelope inside a glass envelope. The inner quartz tube can stand higher temperatures, and allows for the current to arc between the two electrodes exciting a plasma of mercury, metal halide, or high pressure sodium.

HP or **BHP** Horsepower or brake horsepower, a unit of power, equal to roughly 746 watts.

HVAC The abbreviation for heating, ventilation, and air conditioning.

I

Illumination The intensity of light falling on a surface, usually expressed in footcandles.

Impact Isolation Class (IIC) A rating of the degree of isolation of a floor against the transmission of impact noises.

Infiltration The leakage of air through cracks around windows and other building elements.

Intensity Level The intensity of sound at a given location, measured in watts per square meter, or more commonly in dB where the reference level is 10^{-12} watts/meter^2 or 10^{-16} watts/cm^2.

Inverse Square Law A physical principle that states that the intensity of a phenomenon is inversely proportional to the square of the distance from the source to the measuring device. It holds true for point sources of light, and for sound in an open field.

Invert The lowest point of the inside of a drain, pipe, channel, or other liquid-carrying conduit.

Ion Exchange A process of water softening in which calcium and magnesium ions are replaced by sodium ions. This process is also known as the zeolite process.

Ionization Detector A fire detector which detects the products of combustion (POC) even before they are visible to the naked eye.

K

K Factor The thermal conductivity of one square foot of a material per inch of thickness, with a surface temperature difference of one degree F.

Kilowatt (KW) A unit of electric power, equal to 1,000 watts.

KVA A rating for transformers equal to the product of volts and amperes divided by 1,000. The product of the KVA and the power factor gives the power in kilowatts.

L

Latent Heat The heat added to or removed from a substance when it changes its state. See Sensible Heat.

Light Shelf An overhang, either outside or inside or both, that is used with a clerestory to reflect light up onto the ceiling, and reduce direct light adjacent to the window below.

Lumen A unit of light, defined as the amount of light passing through one square foot at a distance of one foot from a one candlepower source.

Luminaire A complete light fixture, including lamps.

M

Mass Law The theoretical law that states that for each doubling of mass in a wall, there is a 6 dB drop in the amount of sound transmitted. In actual practice, it is usually closer to 5 dB.

Mean Radiant Temperature (MRT) The weighted average of all of the temperatures of all of the surfaces visible from a given position.

N

NC Curves A single-number system for specifying a maximum SPL level in a given location, using standardized reference contours. The curves weight the frequencies to which the human ear is sensitive.

NEC The abbreviation for National Electrical Code.

Neutral The wire or conductor in an electrical system that is equidistant in voltage from the phase conductors of the system. It is not the same as ground.

Noise Reduction Coefficient (NRC) A one-number rating system giving the average sound absorption coefficient of a material at frequencies of 250, 500, 1,000 and 2,000 Hz.

O

Occupancy The purpose for which a building is intended to be used.

Occupancy Group A designation for a group of several occupancies that have comparable fire safety considerations, and which are therefore grouped together by the code.

OSHA The abbreviation for the Occupational Safety and Health Act, which regulates working conditions.

P

Passive Solar Design The practice of orienting and sizing a building, its windows, and its internal masses in such a way that it responds to the sun and to the climate, without the use of mechanical equipment.

Perm The unit of permeability for a given material, expressing the resistance of the material to the penetration of water or water vapor through it. One perm is equal to the flow of one grain of water vapor through one square foot of surface area per hour with a pressure difference of one inch of mercury.

Permeability The property of permitting passage of water or water vapor through a material without causing rupture or displacement.

Potable Water Water that is suitable for drinking.

Power Factor (PF) In an electrical circuit, the ratio of real power in watts to the product of voltage and current.

Power Level (PWL) The logarithmic expression for the acoustical power at the source of a sound. It is also known as Acoustic Power Level.

PPM The abbreviation for parts per million.

Psychrometric Chart A graph showing the relationships between temperature, humidity, relative humidity, and enthalpy.

R

Reheat The adding of sensible heat to a supply air stream which has been previously cooled.

Relative Humidity (RH) The ratio of the moisture content of the air to the maximum possible content at the same temperature.

Reverberation The persistence of sound in an enclosed space after the source has stopped.

Reverberation Time The time it takes a 60 dB sound to completely die away in a closed room after the source has stopped.

S

Sabin The unit of sound absorption equivalent to the absorption of one square foot of open window.

Sensible Heat Heat that changes the temperature of a substance, and does not represent the addition of any moisture to the substance, or any change of state. See Latent Heat.

Shading Coefficient (SC)   The ratio of the solar heat gained through a window with ading devices to the solar heat gained by a single pane double strength clear glass window. Shading devices, such as Venetian blinds, lower the SC.

Siamese Fitting   A Y-shaped hose attachment at the base of a building, which allows the fire department to connect a fire hydrant through a pumper truck to provide or augment water flow to a standpipe.

Smoke Developed Rating   A numerical rating derived from a standardized fire test procedure. Larger numbers indicate a greater density of smoke.

Smokeproof Enclosure   A continuous enclosed stairway separated from the building at each floor by an open vestibule that allows smoke to vent away without entering the stair.

Soil   A sanitary drainage term referring to the waste from urinals, water closets, and fixtures of similar function.

Sone   A subjective system of measuring loudness, based on the reference point of one sone equal to a sound pressure level of 40 dB.

Sound Level Meter   A meter that measures the sound pressure level and gives a reading in dB.

Sound Pressure Level (SPL)   The logarithmic expression of the pressure exerted by sound waves on the receiver. The reference pressure is 2×10^{-5} newtons per square meter.

Sound Transmission Class (STC)   A single-number rating for the evaluation of a particular construction cross-section in terms of its transmission of airborne sound. The higher the STC rating, the more effective the construction is at stopping airborne sound.

Sprinkler System   A system used to extinguish fires automatically by releasing water or other substances. See Dry Pipe Sprinkler and Wet Pipe Sprinkler.

Stack Vent   The portion of a soil or waste stack that is above the highest branch drain connected to the stack. Its sole function is to vent to the outside air.

Standpipe   A vertical supply pipe for firefighting. Dry standpipes are empty, and must be connected to a fire hydrant via a siamese connection and a pumper truck. Wet standpipes are pressurized and filled with water, to serve attached hoses within the building, on each floor. Wet standpipes also have siamese connections to allow the water flow to be augmented from fire hydrants.

Star Connection   A method of connecting the windings on a three-phase transformer in which one end of all three windings is connected to a common neutral center point, forming a Y shape. Same as Wye Connection.

Starter   A device that starts the arc in a neon or fluorescent lamp, or a contactor and overload relay used in starting some electric motors.

Steam Trap   A valve that permits passage of air or water, but not steam, often used with steam radiators.

Sweating   The method of soldering copper plumbing, or the condensation of water on cold pipes or building materials.

T

Thermosiphon   The method of using a heated surface and the resulting convection to move a fluid out of a space. In solar hot water heaters, the collector is below the storage tank, and the water is circulated automatically by convection when it is heated. Thermosiphoning may also be used to ventilate a building by sending the warmed air out the top, and siphoning in cooler air at the bottom.

Three Pipe System   A hot and chilled water system having a common return pipe for both supply lines.

Ton   The amount of cooling required to create a ton of ice in a 24 hour period, equal to a steady rate of 12,000 Btuh.

Total Heat   See Enthalpy.

Transmission Loss (TL)   The reduction of sound that occurs when a given wall transmits sound from one room to an adjacent room, expressed in decibels.

Trombe Wall   A form of mass wall that transfers heat by causing convection into the room behind it, as well as conduction.

Two Pipe System   A hot or chilled water system having only a supply and return line. It can supply only heated or chilled water to a zone, but not both simultaneously.

U

UBC   The abbreviation for Uniform Building Code, one of the most widely-adopted model building codes in the United States.

U Factor or U Value   The thermal conductivity of a particular wall section, expressed in Btu's per hour per degree Fahrenheit per square foot.

V

Vacuum Breaker   An automatic valve that admits air into a supply pipe rather than allowing the pipe to suction or siphon polluted water back into the supply.

Variable Air Volume (VAV)   An air conditioning system that accommodates thermal load changes by varying the flow of supply air into a conditioned space instead of varying the temperature of the air.

Vent Stack   A vertical pipe that vents several sanitary drainage lines, in order to break the siphoning suction which would occur when water drops through the system.

W

Watt   The basic unit of electrical power, equal to the product of volts and amperes in direct current systems, equal to 3.41 Btuh.

Wavelength   The length of one complete cycle or waveform, for light or sound waves. In light, the dominant wavelength determines the perceived color.

Wet Bulb Temperature (WB)   The temperature attained by a glass thermometer whose bulb is covered with a wet sock and placed in an air stream moving at 1,000 feet per minute.

Wet Pipe Sprinkler   A sprinkler system that is continually pressurized with water. If a fusible sprinkler opens, water is immediately forced through the sprinkler head.

Wet Standpipe   See Standpipe.

Wind Chill Index   See Chill Factor.

Wye Connection   See Star Connection.

Z

Zeolite   A substance used in water softening, in which a filtering tank is recharged by passing a salt solution through it. The tank may then be used as a filter for the free ions associated with hard water.

Zone   A portion of a building controlled by a single thermostat, because its spaces have similar heating or cooling needs.

BIBLIOGRAPHY

The following list of books is provided for candidates who may wish to do further research or study in Mechanical and Electrical Systems. Most of the books listed below are available in college or technical bookstores, and all would make welcome additions to any architectural bookshelf. In addition to the course material and the volumes listed below, we advise candidates to review regularly the many professional journals, which are available at most architectural offices.

ANSI A117.1
(Handicapped Standards)
American National Standards Institute

Architectural Graphic Standards
Ramsey and Sleeper
Wiley and Sons

ASHRAE Handbook of Fundamentals
American Society of Heating, Refrigerating and Air-Conditioning Engineers

Concepts in Architectural Acoustics
M. David Egan
McGraw-Hill

Design with Climate
Victor Olgyay
Princeton University Press

IES Handbook Reference Volume
John Kaufman
Illuminating Engineering Society

Mechanical and Electrical Equipment for Buildings
Stein/Reynolds/McGuinness
Wiley and Sons

National Electrical Code
National Fire Protection Association

National Plumbing Code
American Society of Mechanical Engineers

The Passive Solar Energy Book
Edward Mazria
Rodale Press

Uniform Building Code
International Conference of Building Officials

Lesson 1

1. **A**   Insolation (solar gain) is a form of radiation. It does not require contact (the earth does not touch the sun), is transmitted through clear objects such as glass, and can be blocked by opaque objects.

2. **C**   Convection is the only heat transfer process in which *direction* (up, down, or sideways) makes any difference, and therefore is the only process in which relative height would make a difference.

3. **B**   The MRT is defined as the weighted average of the surrounding surface temperatures.

4. **A**   The CLTD is based on wall mass, orientation, and color, which relate to conduction and radiation.

5. **D**   The resistance determines the temperature gradient and the conduction, and is determined by the conductivity.

6. **C**   A degree day measures the temperature over a period of time against a reference temperature (65°F). A design day is a theoretical day which is "worse" (either colder or hotter) than 98 percent of the days experienced at a site, and allows us to size the plant to cover most days.

7. **D**   See the answer to the above question and page 11.

8. **C**   A and B are immediately eliminated because q_{CLTD} is always a heat gain. D contains q_r, which is a gain from the sun. C contains *only* heat *losses*.

9. **A**   Design day temperatures can only refer to a particular moment in time, eliminating B and D. Temperature does not measure latent heat, eliminating C. Design days are expressly used for equipment sizing.

10. **D**   The evaporation of steam is a prime example of latent heat transfer. In addition, heat is conducted into the pot from whatever it is sitting on, and the heat is drawn away by the convective flow in the air around the pot (and under the pot, if it is a gas stove).

Lesson 2

1. **A**   The ambient air temperature must make up for the radiant loss to the surroundings.

2. **D**   The *relative* humidity changes with temperature. All of the other choices are absolute measures of the amount of water in the air, and say nothing about how close the air is to saturation.

3. **C**   A sling psychrometer is an instrument designed to measure the wet bulb temperature. This is compared with the dry bulb temperature to determine the relative humidity.

4. **C**   There are several kinds of thermal storage walls, but a Trombe wall is distinguished by the fact that it causes a convective loop.

5. **D**   Hot arid climates often experience a large diurnal temperature swing, so damping or averaging temperature swings is an excellent strategy. If there is a water supply, there will be a cooling effect from evaporation. Unfortunately, if the outside air is hot, there is little benefit from ventilation. This is especially true if the other two strategies succeed in cooling, since the inside air will then be cooler than the outside air. (Note: Hot *humid* climates are more suited for ventilation, where that is often the only passive possibility.)

6. **C**   The declination angle is affected only by the time of year, not the time of day.

7. **A** Refer to the psychrometric chart. Find where the 60°F vertical line crosses the 70% RH curved line and follow the diagonal line upward and to the left to read 54°F.

8. **A** There is no change in enthalpy due solely to air movement.

9. **A** The earth stores heat lost into it, and changes temperature much more slowly than outside air. However, earth is not a good insulator, moisture usually causes problems, and the weight of the earth increases structural costs.

10. **D** Since focusing collectors must track the sun, they are usually more difficult to mount than flat plate collectors.

Lesson 3

1. **C** The four-pipe system can bring the heating and cooling capacity to the coils in the fan coil system.

2. **B** Systems which use the compressive refrigeration cycle reverse the flow from hotter to cooler objects (C). A heat pump uses the refrigeration cycle (A). An evaporative chiller, such as a swamp cooler, can do the same (D), but is most often used for increasing the flow rate of heat away from the condenser coils on a compressive refrigeration cycle. The only incorrect choice is B.

3. **D** The electric reheat system (I) normally cools, but can heat when the resistance coil is on. The double duct system (II) distributes only air and can heat or cool, as needed, at each mixing box. The fan coil system (III) can also heat or cool at each unit.

4. **A** The evaporative chiller uses the latent heat of evaporation to transfer heat to the outside air.

5. **A** Both the unitary system and the heat pump system can have a meter on each unit, thus keeping track of individual usage and permitting separate billing.

6. **C** The heat pump system uses a heat sink connected to several heat pumps. When one zone is cooling at the same time as another is heating, for example in the morning if one zone is in the easterly part of a building and another is in the west, then the system can balance itself without a boiler or chiller. This can happen fairly often in mild climates.

7. **C** The condenser is the high pressure coil, which is hottest and gives off the most heat.

8. **B** The one pipe system is run in series, while the others are parallel systems.

9. **C** The one and two pipe systems provide either heating or cooling, but not both simultaneously. The three and four pipe systems can provide both simultaneously.

10. **A** It is called return air because it is *returning* to the *plant*.

Lesson 4

1. **D** The "warm" colors come from the *lower* Kelvin temperatures. Psychologically warm colors happen to be generated by lower temperatures than the psychologically cool colors.

2. **C** The retina is the sensing surface. The lens focuses, the iris adjusts the amount of light, and there are no cilia in the eye (there is a ciliary muscle).

3. **B** Glare comes from high contrast. It can occur at low light levels, for example from headlight glare at night. It can come directly from a source, not just from reflections, for example by looking directly at the sun.

4. **B** Translucent means that light can pass through, but not an image.

5. **C** Ambient lighting is usually diffuse, and diffuse lighting does not cause veiling reflections, so it is a good light for reading flat surfaces. Mirrors are not diffusing, but specular. A flat paint would be a diffuse reflecting surface.

6. **C** A footcandle is a measure of illumination. Candlepower measures the power at the source, a lumen measures flux, and a footLambert measures reflected or transmitted brightness.

7. **A** According to the inverse square law, the illumination at 20 feet is $(10/20)^2$ that at 10 feet, or 1/4 of 24 fc, or 6 fc.

8. **C** Low voltage lamps have smaller filaments, and therefore allow more precise optical control of light placement.

9. **D** An R-40 bulb is $40 \times 1/8"$ or 40/8 or 5" in diameter.

10. **D** All three statements are true.

Lesson 5

1. **A** $I = V/R = > 120 \text{ v}/8 \ \Omega = 15 \text{ amps}$

2. **A** $P = V \times I \times PF = > 120 \times 15 \times 1$
$= 1,800 \text{ watts}$

3. **D** $R = 1/(1R_1 + 1/R_2 + 1/R_3)$
$= 1/(1/4 + 1/4 + 1/2)$
$= 1/1$
$= 1.0 \text{ ohms}$

4. **B** $V_2 = V_1/\sqrt{3}$
$V_1 = V_2 \times \sqrt{3} = 120 \text{ v} \times 1.73 = 208 \text{ v}$

5. **C** There are three wires from the secondary of a three wire transformer: a line, the neutral, and the opposite line. The secondary winding has two options for connection: one line to the other line, or either line to neutral.

7. **A** Universal motors are the typical motors in small appliances, allowing variable speeds. See page 97.

8. **D** Armored cable may not be embedded in concrete.

9. **D** A ground fault interrupter is specifically designed to pick up low level electrical short circuits, which often occur with no equipment switched "on," and then shut the entire circuit off.

10. **C** A *demand* surcharge is based on the peak usage, or demand, and encourages the customer to reduce those peaks. Reduced peaks reduce the need for new power generating plants.

Lesson 6

1. **C** Noise criteria or NC curves, see page 116.

2. **A** IL or intensity level, see page 107.

3. **B** The sound transmission class (STC) is a widely accepted method of rating walls and doors according to their typical or overall resistance to sound transmission. See page 119.

4. **A** The reverberation time is defined as the time required for a sound to decay 60 dB in a space. The return time for an echo depends on the distance to the reflecting surface, and is only vaguely related to reverberation time. Reverberation time is *shorter* in a dead space.

5. **C** 40 dB is acceptable for a hospital or a quiet office. 10 dB is not a practical level, and 70 dB is definitely too loud.

6. **A** Doubling the distance results in a decrease of 6 dB.

7. **D** Doubling the source power results in an increase of 3 dB.

8. **A**   Doubling the mass theoretically results in a 6 dB drop.

9. **D**   Carpeting and drapes are much more absorptive surfaces than concrete. Increasing the absorptivity decreases both the reverberation time and the volume.

10. **C**   The STC is comparatively high, so the wall is good at stopping transmitted sound. The STC says nothing about absorptivity or reflected sound.

Lesson 7

1. **C**   Rain water is slightly acidic and tends to corrode metal pipes. Furthermore, contaminants may be absorbed into the rain water, which may make it unhealthful, while ground water tends to be partially filtered of such contaminants.

2. **C**   When water seeps into the ground, it dissolves minerals and becomes hard. This is not usually hazardous to humans, but tends to clog pipes. Hardness of water is not the same as acidity; they are different problems and are treated differently.

3. **D**   Acidic water causes corrosion, not deposition. Where water is acidic, corrosion-resistant materials such as copper and PVC are appropriate for piping. Hardness, not acidity, can be controlled by the zeolite process. Plastic pipe, including PVC, should never be used in exposed locations above ground because it deteriorates when exposed to ultra-violet light, which is present in direct sunlight.

4. **A**   The lower the pH reading, the more acidic the water. 5 is very low for water. A pH of 7 is neutral. There is no such thing as pH 17.

5. **B**   Copper is soldered, which relies on heat to melt the flux and solder.

6. **D**   100 psi − 10 psi = 90 psi at the building. The faucet requires 10 psi, leaving 90 − 10 = 80 psi to lift the water within the building. 80 psi × 2.3 ft/psi = 184 feet.

7. **B**   II only. I is true of both the downfeed and the pneumatic systems. II is true only of the downfeed. III is not true.

8. **B**   II only. I is not true. II is true, and is the reason for separation. III is not true; in general, neither storm drainage nor sanitary drainage is pressurized.

9. **D**   The water in the trap keeps the methane gas that is generated in the sewers below from rising up into the building.

10. **A**   A catch basin is an inlet at the low point of a swale or other drainage collection area to collect surface runoff. (An interceptor collects grease, etc.)

Lesson 8

1. **B**   Compartmentation attempts to contain the fire within the area in which it started for a given time, so that occupants may safely escape the building, damage spread may be limited, and the fire department may combat the blaze.

2. **C**   44 inches is the *minimum* exit corridor width.

3. **C**   The maximum distance from a room to an exit stairway is 150 feet unless the building is sprinklered. See page 150.

4. **B**   The hose is 100 feet long, and every point on the floor must be reached by a 30-foot spray from that hose.

5. **A**   Hot air and smoke from small or smoldering fires often heat the ceiling evenly but slowly. When the temperature finally goes above ignition, it does so for nearly the entire ceiling at once. This is called flashover.

6. **C** Occupancy groups are defined by function. Construction materials determine the building construction type.

7. **C** The fire department must pump water into a dry standpipe to use it for fighting fires. Building occupants have no way to use them, unlike wet standpipes. Halon gas or carbon dioxide may fill an extinguisher system similar to a sprinkler system, but not a standpipe. There is no special drainage system provided for fighting fires.

8. **A** Sprinklers function whether or not someone is there to observe the fire. They are more expensive than the standpipes, they

can damage artwork, books or records, and they may *not* be used as part of the domestic plumbing.

9. **A** The ionization detector will detect invisible products of combustion. Smoke detectors only detect smoke of significant density. A fusible link only melts when it gets hot enough, which may be after some other part of the space is already well involved in the fire. The same is true of a heat detector.

10. **D** All three are required to have a siamese connection at the base, so that the fire department can add water to the system.

EXAMINATION DIRECTIONS

The examination on the following pages should be taken when you have completed your study of all the lessons in this course. It is designed to simulate the Mechanical and Electrical Systems division of the Architect Registration Examination. Many questions are intentionally difficult in order to reflect the pattern of questions you may expect to encounter on the actual examination.

You will also notice that the subject matter for several questions has not been covered in the course material. This situation is inevitable and, thus, should provide you with practice in making an educated guess. Other questions may appear ambiguous, trivial, or simply unfair. This too, unfortunately, reflects the actual experience of the exam and should prepare you for the worst you may encounter.

Answers and complete explanations will be found on the pages following the examination, to permit self-grading. **Do not look at these answers until you have completed the entire exam.** Once the examination is completed and graded, your weaknesses will be revealed, and you are urged to do further study in those areas.

Please observe the following directions:

1. The examination is closed book; please do not use any reference material.
2. Allow about 60 minutes to answer all questions. Time is definitely a factor to be seriously considered.
3. Read all questions *carefully* and mark the appropriate answer on the answer sheet provided.
4. Answer all questions, even if you must guess. Do not leave any questions unanswered.
5. If time allows, review your answers, but do not arbitrarily change any answer.
6. Turn to the answers only after you have completed the entire examination.

GOOD LUCK!

EXAMINATION ANSWER SHEET

Directions: Read each question and its lettered answers. When you have decided which answer is correct, blacken the corresponding space on this sheet. After completing the exam, you may grade yourself; complete answers and explanations will be found on the pages following the examination.

1 Ⓐ Ⓑ Ⓒ Ⓓ
2 Ⓐ Ⓑ Ⓒ Ⓓ
3 Ⓐ Ⓑ Ⓒ Ⓓ
4 Ⓐ Ⓑ Ⓒ Ⓓ
5 Ⓐ Ⓑ Ⓒ Ⓓ
6 Ⓐ Ⓑ Ⓒ Ⓓ
7 Ⓐ Ⓑ Ⓒ Ⓓ
8 Ⓐ Ⓑ Ⓒ Ⓓ
9 Ⓐ Ⓑ Ⓒ Ⓓ
10 Ⓐ Ⓑ Ⓒ Ⓓ
11 Ⓐ Ⓑ Ⓒ Ⓓ
12 Ⓐ Ⓑ Ⓒ Ⓓ
13 Ⓐ Ⓑ Ⓒ Ⓓ
14 Ⓐ Ⓑ Ⓒ Ⓓ
15 Ⓐ Ⓑ Ⓒ Ⓓ
16 Ⓐ Ⓑ Ⓒ Ⓓ
17 Ⓐ Ⓑ Ⓒ Ⓓ
18 Ⓐ Ⓑ Ⓒ Ⓓ
19 Ⓐ Ⓑ Ⓒ Ⓓ
20 Ⓐ Ⓑ Ⓒ Ⓓ
21 Ⓐ Ⓑ Ⓒ Ⓓ
22 Ⓐ Ⓑ Ⓒ Ⓓ
23 Ⓐ Ⓑ Ⓒ Ⓓ
24 Ⓐ Ⓑ Ⓒ Ⓓ
25 Ⓐ Ⓑ Ⓒ Ⓓ

26 Ⓐ Ⓑ Ⓒ Ⓓ
27 Ⓐ Ⓑ Ⓒ Ⓓ
28 Ⓐ Ⓑ Ⓒ Ⓓ
29 Ⓐ Ⓑ Ⓒ Ⓓ
30 Ⓐ Ⓑ Ⓒ Ⓓ
31 Ⓐ Ⓑ Ⓒ Ⓓ
32 Ⓐ Ⓑ Ⓒ Ⓓ
33 Ⓐ Ⓑ Ⓒ Ⓓ
34 Ⓐ Ⓑ Ⓒ Ⓓ
35 Ⓐ Ⓑ Ⓒ Ⓓ
36 Ⓐ Ⓑ Ⓒ Ⓓ
37 Ⓐ Ⓑ Ⓒ Ⓓ
38 Ⓐ Ⓑ Ⓒ Ⓓ
39 Ⓐ Ⓑ Ⓒ Ⓓ
40 Ⓐ Ⓑ Ⓒ Ⓓ
41 Ⓐ Ⓑ Ⓒ Ⓓ
42 Ⓐ Ⓑ Ⓒ Ⓓ
43 Ⓐ Ⓑ Ⓒ Ⓓ
44 Ⓐ Ⓑ Ⓒ Ⓓ
45 Ⓐ Ⓑ Ⓒ Ⓓ
46 Ⓐ Ⓑ Ⓒ Ⓓ
47 Ⓐ Ⓑ Ⓒ Ⓓ
48 Ⓐ Ⓑ Ⓒ Ⓓ
49 Ⓐ Ⓑ Ⓒ Ⓓ
50 Ⓐ Ⓑ Ⓒ Ⓓ
51 Ⓐ Ⓑ Ⓒ Ⓓ
52 Ⓐ Ⓑ Ⓒ Ⓓ
53 Ⓐ Ⓑ Ⓒ Ⓓ

FINAL EXAMINATION

1. Which of the following does an open fireplace contribute to the environment?

   I.   Radiant heat

   II. Convection

   III. Latent heat

   A. I only   **C.** III only

   B. II only   **D.** I and II

2. Given a coefficient of utilization for a given room of .54, a maintenance factor of .8, and a typical four lamp fixture with 2900 lumens/lamp, how many fixtures are needed to provide a light level of 50 footcandles if the room size is 100 × 100 feet?

   A. 10 fixtures   **C.** 60 fixtures

   B. 30 fixtures   **D.** 100 fixtures

3. Which of the following strategies reduces overall energy consumption in a building?

   I.   An HVAC system with an economizer cycle

   II. A daylighting design which provides light but little heat

   III. A wet standpipe system

   A. I and II   **C.** I and III

   B. II and III   **D.** I, II, and III

4. Which of the following plumbing devices most affects the safety and health of a building's occupants?

   A. A gate valve

   B. A globe valve

   C. A trap

   D. An interceptor

5. Which of the following is true?

   A. When DB equals WB, RH equals 100%.

   B. DB cannot equal WB.

   C. DB cannot be greater than WB.

   D. RH can exceed 100%.

6. Which of the following is NOT usually true of a passive solar design?

   A. It blocks the sun in overheated months.

   B. It stores heat in thermal mass or phase change materials.

   C. It is cheaper to build than a conventional building.

   D. It relies on a backup system for dealing with extremes in weather.

7. Which of the following statements is NOT true of a heat pump?

   A. It works in the same manner as an air conditioner.

   B. It cannot deliver more heat energy than the energy required to run the pump.

   C. It works only between certain temperature ranges.

   D. It is often used in conjunction with a recirculating heat sink.

8. What is the primary purpose of a leach field?

   A. To store storm runoff until it can seep into the ground.

   B. To store sanitary drainage until it can be pumped out and removed.

   C. To remove greywater before it enters the public sewage system.

   D. To allow sewage cleared of solid matter to seep into the ground.

9. Reverberation time depends on which of the following factors?

   I.   Room volume

   II. Sound volume

   III. Absorptivity of surfaces

   A. I and II   **C.** I and III

   B. II and III   **D.** I, II, and III

10. One of the following is an appropriate lighting design. Which is it?

   A. A fluorescent cove light, casting indirect light on the ceiling, is used to bring out the sparkle in a display of diamonds.

   B. A mercury vapor lamp is used in an operating room to get the highest possible light levels and proper color rendition of the patient.

   C. Suspended incandescent globes are used in a computer room to reduce the sterile atmosphere and provide warm colors that do not reflect in the video screens.

   D. Low voltage lamps are used in lighting a display because of their tight focus.

11. Which of the following are included in the refrigeration cycle?

   I.   Freon

   II. Compressor

   III. Expansion valve

   A. I and II   **C.** I and III

   B. II and III   **D.** I, II, and III

12. Rank the following walls in the order of their ability to stop the transmission of sound.

   I.   A concrete wall

   II. A stud wall with plaster on both faces

   III. A stud wall with plaster and an extra layer of acoustical tile on both faces

   A. III, II, I   **C.** II, III, I

   B. I, III, II   **D.** III, I, II

13. Rank the following materials in the order of their ability to absorb sound.

   I.   Carpet

   II. Plaster

   III. Acoustical tile

   A. III, II, I   **C.** II, III, I

   B. I, III, II   **D.** III, I, II

14. Which of the following is UNRELATED to the definition of degree day?

   A. History of outside temperature

   B. Design day temperature

   C. Reference temperature

   D. Number of hours per day at a given temperature

15. What is the purpose of the trap in a sewage system?

   A. To keep sewer gas from passing up into the building

   B. To catch small valuables and other materials before they pass down the drain

   C. To catch grease before it clogs the drain

   D. To provide a cleanout when drains become clogged

16. Which of the following is NOT one of the purposes of a vent stack?

    A. To provide a vent for sewer gases

    B. To keep the water in the trap from getting siphoned out

    C. To vent several different drains

    D. To provide a vent for the bathroom and its odors

17. Which of the following is NOT a reasonable strategy for a building located in a hot arid climate?

    A. Provide maximum ventilation to the outside to dissipate internal heat gain.

    B. Provide high thermal mass walls.

    C. Provide an internal courtyard with shading devices and/or evaporation sources.

    D. Place homes in close proximity for mutual shading.

18. Which of the following are NOT fire extinguishing agents?

    I.   Halon

    II. Freon

    III. Carbon Dioxide (CO_2)

    IV. Metal Halide

    A. I and II   **C.** II and III

    B. III and IV   **D.** II and IV

19. Given a 20 × 30 × 10 foot room, with concrete floor, walls and ceiling, what is the total absorptivity of the room, if the absorption coefficient for concrete is 0.02 sabins?

    A. 24 sabins   **C.** 84 sabins

    B. 44 sabins   **D.** 120 sabins

20. Which of the following is NOT a high intensity discharge lamp?

    A. Tungsten halogen

    B. High pressure sodium

    C. Mercury vapor

    D. Metal halide

21. If one Evinrude outboard motor results in 70 dB of sound in the boat, what is the sound level of two motors?

    A. 64 dB   **C.** 76 dB

    B. 73 dB   **D.** 140 dB

22. If an outboard motor results in a sound intensity level of 60 dB at a distance of 40 feet, what is the IL at 80 feet?

    A. 30 dB   **C.** 54 dB

    B. 40 dB   **D.** 57 dB

23. A proposed building has a peak cooling load of 144,000 Btuh for the design day. How many tons of air conditioning does this represent?

    A. 12 tons   **C.** 144 tons

    C. 72 tons   **D.** 288 tons

24. What is the function of a catch basin?

    A. It holds overflow until it can be drained away later.

    B. It collects the water under a fixture (it is often called a sink).

    C. It collects surface runoff and admits it to a storm drainage line.

    D. It intercepts grease and solid objects before they clog a main sewer line.

25. If a wall has a total thermal resistance of 21.6 s.f. °F/Btuh (including air films), what is the U value of the wall?

    A. 0.046 Btuh / ft² °F

    B. 0.46 Btuh / ft² °F

    C. 2.16 Btuh / ft² °F

    D. 21.6 Btuh / ft² °F

26. What is the likely result of excessively hard water in piping?

    A. Steel pipe will corrode.

    B. Plastic pipe will fill with encrustations.

    C. Copper pipe will rust.

    D. Steel pipe will plug up with depositions.

27. What is the likely result of excessively acidic water in piping?

    A. Steel pipe will corrode.

    B. Plastic pipe will fill with encrustations.

    C. Copper pipe will rust.

    D. Steel pipe will plug up with depositions.

28. What is the special danger of a class C fire?

    A. That there will be a chemical reaction between the burning objects and the extinguisher medium.

    B. That the user of the extinguisher may be electrocuted.

    C. That the fire will spread more rapidly than any other fire.

    D. There is no special danger, since class C is the least hazardous type of fire.

29. Which of the following statements is correct?

    A. A dry pipe sprinkler system causes less water damage when discharging and is therefore preferable to a wet pipe system.

    B. A POC detector responds only to invisible gases.

    C. A fire alarm should be both audible and visible.

    D. A dry standpipe should be connected to a dependable water source at all times.

30. In remodeling an existing school building which has a steam heat system and no air conditioning, the intent is to create separate living units. Which of the following systems is probably the best choice for the building?

    A. A double duct forced air system, which would provide each unit either heating or cooling, independent of the other units.

    B. A variable air volume system, which would be the most efficient, even if one unit is being heated while an adjacent unit is being cooled.

    C. A heat pump system, which would not require the addition of ductwork, and would allow each unit to be heated or cooled, as desired.

    D. A multizone system, which would provide for different temperature zones with less space required than any of the others.

31. Which of the following systems includes an evaporator coil and a condenser coil?

   I. Fan coil system

   II. Heat pump system

   III. Refrigeration cycle

   A. I and II   **C.** I and III

   B. II and III   **D.** I, II, and III

32. What system or equipment uses only evaporation and no compression to cool water?

   A. Double duct system

   B. Fan coil system

   C. Unitary system

   D. Evaporative chiller

33. Which of the following terms means brightness?

   A. Luminance   **C.** Glare

   B. Illumination   **D.** Radiation

34. In designing a hotel, which of the following must be provided in at least some of the bathrooms to make them accessible to the handicapped?

   I. Grab bars

   II. Shower stall seat

   III. Shower stall large enough to accommodate a five-foot turning circle

   A. I and II   **C.** I and III

   B. II and III   **D.** I, II, and III

35. A single duct constant volume system with electric reheat has all the following characteristics, EXCEPT

   A. its initial cost is low.

   B. it can heat or cool, but not both simultaneously.

   C. it is inefficient if there is very much need for heating.

   D. it includes an electric resistance heater adjacent to the cooling coil.

36. A fitting outside a building which provides two or more connections through which water can be pumped to a standpipe or sprinkler system is called a

   A. hydrant.

   B. ball drip.

   C. siamese fitting.

   D. combination standpipe.

37. The intensity of a light source is measured in what units?

   A. Lumens   **C.** Footcandles

   B. Candlepower   **D.** Lamberts

38. What system removes heat from the air exhausted from a building and transfers it to the incoming fresh air?

   A. Fan coil system

   B. Heat pump system

   C. Air-to-air heat exchanger

   D. Variable air volume system

39. What is the thermal process which takes place in a fluid medium?

   A. Radiation   **C.** Convection

   B. Conduction   **D.** Transmission

40. The ability of a surface to emit heat by radiation relative to that of a black body at the same temperature is called
   A. radiation.　　C. transmissivity.
   B. emissivity.　　D. reflectivity.

41. Which system always has both heated air and chilled air running through it?
   A. Double duct system
   B. Multizone system
   C. Fan coil system
   D. Variable air volume system

42. A stack vent is
   A. the same as a vent stack.
   B. the air intake line for all the fixtures.
   C. a pipe into which all the soil and waste lines empty.
   D. the extension of a soil stack to the outdoors above the highest branch drain.

43. What factor affects both reverberation time and loudness in enclosed spaces?
   A. Acoustical intensity　　C. Absorptivity
   B. Transmissivity　　D. Frequency

44. Insolation is
   A. a material having high resistance to heat flow.
   B. heat gain from the sun.
   C. heat energy transmitted by radiation.
   D. thermal resistance.

45. Pitch as perceived by the ear is determined by a sound's
   A. frequency.　　C. intensity.
   B. velocity.　　D. pressure.

46. CLTD (cooling load temperature differential) includes the effects of all of the following, EXCEPT
   A. time lag.　　C. conduction.
   B. radiation.　　D. humidity.

47. Both sound intensity level (IL) and sound pressure level (SPL) are measured in
   A. watts.　　C. sones.
   B. decibels.　　D. hertz.

48. Which of the following systems can be used if each zone is to be billed separately?
   I.   Heat pump system
   II. Multizone system
   III. Unitary system
   A. I and II　　C. I and III
   B. II and III　　D. I, II, and III

49. A step down transformer in a building is used to
   A. decrease the voltage.
   B. decrease the current.
   C. decrease the power.
   D. convert alternating current to direct current.

50. The ratio of the actual amount of water vapor in the air to the maximum amount that the air could contain at the same temperature is called the
   A. humidity ratio.
   B. effective humidity.
   C. enthalpy.
   D. relative humidity.

51. The *Natural Step* is an approach to the environment that follows which of the following principles?

    I.   The biosphere affecting humans is a relatively stable and resilient zone which includes five miles into the earth's crust and five miles into the troposphere.

    II. Improved technologies have dramatically increased the number and quantity of available natural resources.

    III. Toxic substances released into either the sea or atmosphere will only influence areas adjacent to the toxic source.

    IV. Using building materials that are recycled is an adequate sustainable design approach.

    A. I

    B. II

    C. II, IV

    D. None of the above

52. Which of the following is a consultant that might be employed in a sustainable design project?

    I.   Wetlands engineer

    II. Energy commissioner

    III. Landscape architect

    IV. Energy modeling engineer

    A. I

    B. I, II

    C. I, III, IV

    D. All of the above

53. Sustainably designed architecture requires attention to which of the following building elements?

    I.   Solar shading devices

    II. Urban heat island effect

    III. Increased parking

    IV. Fenestration and glazing

    A. I, II, IV

    B. I, IV

    C. I, II

    D. All of the above

The examination answers and explanations will be found on the following pages.

Do not look at the answers until you have completed the exam.

EXAMINATION ANSWERS

1. **D** There is radiant heat from the hot gases and the embers (I), and the gases themselves rise rapidly (convection). Even the warm fireplace causes the air next to it to rise (II). However, little or no evaporation takes place outside of the combustion itself (III is incorrect).

2. **D** The formula for the illumination in a room is:

   E = (fixtures × lamps/fixture × lumens/lamp × maintenance factor × CU)/area

   Solving for the number of fixtures, given the desired illumination:

   Fixtures = (area × E)/(lamps/fixture × lumens/lamp × maintenance factor × CU)

   Fixtures = (10,000 s.f. × 50 fc) / (4 lamps/fixture × 2,900 lumens/lamp × .8 × .54)

   = 99.77 fixtures, which we round up to 100 fixtures

3. **A** An economizer cycle shuts off the refrigeration and uses outside air when the temperatures are low enough, thus reducing the plant load (I). Daylighting reduces the need for artificial lighting, which reduces the energy required. However, daylighting may also increase the heat gain and thus the cooling load, but that is not the case here, since we are told that the design provides little heat (II). A standpipe system is a fire safety device which does nothing until there is a fire, and thus has little to do with energy consumption (III).

4. **C** The trap is the device which keeps the sewer gas from entering the building, which would be harmful to the occupants. The gate and globe valves (A and B) simply regulate water flow. The interceptor (D) keeps solid objects and grease out of the sewer system, which affects the sewage treatment plant, but not the occupants.

5. **A** When the dry bulb temperature DB (not the decibel level, which is dB) is equal to the wet bulb temperature WB, then there is no evaporation cooling the sock on the psychrometer. This means that the relative humidity RH is 100% (A is correct, B is incorrect). When the RH is less than 100%, DB is greater than WB (C is incorrect). Finally, the relative humidity can never be greater than 100%, by definition (D is also incorrect).

6. **C** While passive solar designs often have lower operating costs, their first cost is generally greater, because of the required thermal mass and insulation. In passive solar design, the sun is blocked in overheated months (A), and thermal mass or phase change materials are used to store solar energy without large temperature swings (B). However, a backup system is still necessary (D), because extreme storm conditions are almost certain to occur sometime during the life of the building.

7. **B** A heat pump *does* deliver more energy than the energy required to run it. That is the whole point of the system. It actually moves heat from one point to another, rather than creating heat. It operates on the same cycle as an air conditioner (A), only with the inside and outside reversed. If the temperature range goes beyond the rated range of the refrigerant, the system won't work (C). Several heat pumps will often operate in conjunction with the same heat sink, recirculated within a large building or complex (D).

8. **D** A leach field lets the liquid sewage from a building seep into the ground. Surface runoff should be handled separately, since it would overload the system and back it up (A). A septic tank may need occasional emptying, but a leach field should not store anything (B). Greywater (waste from

fixtures other than toilets and urinals) is typically recycled, by sedimentation and filtering, and hence does not enter the leach field at all (C).

9. **C**   Reverberation time is defined as the time it takes a 60 dB sound to die out after the source has stopped. It depends on room volume and surface absorptivity (I and III). It may vary with pitch, but only because absorptivity varies with pitch. However, it does not vary with sound volume (II).

10. **D**   Low voltage lamps have smaller filaments, and therefore have more tightly controlled optics. This allows a more precise aim, which is useful in highlighting specific objects. A fluorescent cove light provides diffuse ambient illumination, ideal for reading or lighting flat surfaces evenly, but deadly dull in terms of sparkle (A). A mercury vapor lamp provides terrible color renditions of skin and flesh, even with very high light levels. This would be inappropriate or even dangerous in an operating room (B). Suspended incandescent globes provide warm colors and reduce the sterile atmosphere in a computer room, but they also cause glare spots all over the screens, making work very difficult (C). The fluorescent cove light at a low illumination level in choice A would be appropriate for the computer room.

11. **D**   All three are included in the refrigeration cycle. See the diagram on page 54.

12. **B**   Mass is the most important factor in stopping the transmission of sound, and the concrete wall has the greatest mass. Comparing the two stud walls, the acoustical tile will reduce *reflection*, but have minimal effect on *transmission*. However, since there is a slight increase in the amount of material present, the wall with the tile will represent a small improvement. The greatest benefit of the acoustical tile will be in the reduction of the sound level in the room containing the source, and the reduction of reflected sound in the receiving room

13. **D**   The ability of a material to absorb sound is measured by its noise reduction coefficient (NRC). Acoustical tile has a NRC which ranges from .65 to .95, carpet ranges from .3 to .55, and plaster ranges from .02 to .05.

14. **B**   A degree day is defined as the amount by which the average outdoor temperature at a particular location is below the reference temperature of 65°F for one day. Degree days may be summed and stated for a month or a year. Thus, A, C, and D all relate to the definition of degree day. The design day temperature (B) is a temperature which is lower than the outdoor temperature experienced at a given location on 98 percent of the days. It is therefore not directly related to the concept of degree day.

15. **A**   The purpose of the trap is to keep methane gas generated by decomposition from entering the building. It is true that the trap may catch small valuables before they pass down the drain, but this is not the purpose of the trap. The trap catches grease, but this must be removed or dissolved, in order not to clog the trap. The trap may be removed when the drain is clogged, but this is a messy process, and the trap does not usually have a cleanout plug.

16. **D**   The vent stack does not vent the bathroom itself, nor the odors which may occur there. It does vent sewer gases (A), and keep the water from getting siphoned out of the traps for several different drains (B and C).

17. **A**   The outside air in a hot arid climate is too hot and dry during the day, and good design therefore limits air exchange with the

outside. High thermal mass limits temperature swings, storing the heat gained during the day and reradiating it at night (B). A shaded courtyard, especially with some source of evaporative cooling, provides a welcome relief to outside conditions (C). Mutual building shading is often beneficial (D), and many North African cities consist of a collection of contiguous mud brick buildings, with as little sun exposure as possible.

18. **D** Freon (II) is a refrigerant, and metal halide (IV) is a type of high intensity discharge lamp. Carbon dioxide and halon are both used to smother fires without damaging the building's contents.

19. **B** The absorptivity of the room is the surface area (not the volume) multiplied by the absorptivity of the surfaces.

$$A = \text{area} \times \alpha = [(2 \times (30 \times 20)) + (2 \times (30 \times 10)) + (2 \times (20 \times 10))] \times 0.02$$
$$= 44 \text{ sabins}$$

20. **A** Tungsten halogen is an incandescent lamp.

21. **B** Doubling the volume causes an increase of 3 dB. 70 dB + 3 dB = 73 dB.

22. **C** Doubling the distance between source and receiver is equivalent to reducing the sound by one quarter, which results in a drop of 6 dB. 60 dB − 6 dB = 54 dB.

23. **A** One ton is 12,000 Btuh. Therefore, 144,000/12,000 = 12 tons.

24. **C** A catch basin admits water from a swale to a storm drainage line.

25. **A** The U value is the reciprocal of the sum of the resistances. 1/21.6 = .046 Btuh/ft² °F.

26. **D** Hard water means that there are minerals, typically calcium, dissolved in the water. These tend to deposit out in steel piping, clogging it. Corrosion is caused by acidic water (A), plastic pipe is not as

subject to deposition as steel pipe (B), and copper pipe never rusts (C).

27. **A** See the explanation for question 26.

28. **B** A class C fire is an electrical fire, and there could be a short circuit which would conduct through water if water were used to put it out.

29. **C** A fire alarm should be both audible and visible, so that blind or deaf people have some chance of perceiving it. A dry sprinkler system floods with water when it is set off (A). A POC detector (products of combustion) will respond to both visible and invisible gases (B). A dry standpipe is not connected to a water source until a fire occurs, when the fire department connects it to a hydrant via a pumper (D).

30. **C** There are already pipes for the steam heat system. These may be used for the heat sink if they are in good condition, or they may be replaced by new pipes, which would be approximately the same size. All of the other systems are forced air systems which would require major modifications because the required ducting would occupy much more space. The VAV system (B) would require the least, but is not flexible enough to handle different units in different ways. To achieve that flexibility would require a double duct system, with two complete sets of ductwork (A), or a multizone system, with one separate branch for each unit (D). The heat pump system also allows each unit to be billed separately, based on the electricity used by that unit's heat pump.

31. **B** The heat pump system and the refrigeration cycle both have an evaporator coil and a condenser coil.

32. **D** The evaporative chiller, often called a cooling tower, uses only evaporation to cool water. Compression systems use

evaporation and condensation in a closed loop.

33. **A**   Luminance means brightness. Illumination is the amount of light arriving at a surface, not leaving it. Glare results from extreme contrasts in brightness. Radiation is a method of transfer of heat or other energy.

34. **A**   Grab bars are required in all accessible shower stalls, and a seat is required in transfer type shower stalls. But neither a transfer type nor a roll-in type shower stall is required to be large enough to accommodate a five-foot turning circle.

35. **D**   All of the choices are characteristics of the single duct constant volume system with electric reheat, except D. An electric resistance heater is placed in the duct upstream of each diffuser, not adjacent to the cooling coil, to reheat the cooled air if heating is necessary.

36. **C**   A siamese fitting is one located close to the ground outside a building which provides two or more connections to permit the fire department to pump water to a sprinkler system or standpipe.

37. **B**   The intensity of a light source is measured in candlepower (B). The rate of flow of light is measured in lumens (A). A footcandle is a unit of illumination equal to one lumen per square foot (C). A lambert (D) is a unit of luminance or brightness.

38. **C**   The air-to-air heat exchanger removes heat from the exhaust air and warms the incoming fresh air, often at a very high efficiency.

39. **C**   Convection occurs in a fluid medium, often air or water.

40. **B**   Emissivity is the ability of a surface to emit heat by radiation relative to that of a black body at the same temperature. Black surfaces have higher emissivities than white or shiny surfaces.

41. **A**   A double duct system consists of two ducts, one carrying heated air and one carrying cooled air. The amount of air drawn from each duct at each room is controlled by dampers and mixed in a mixing box. The multizone system is similar, except that the mixing boxes are in the mechanical equipment room, and pre-mixed air at the desired temperature is sent out to each zone.

42. **D**   A stack vent is the extension of a soil stack to the outdoors above the highest branch drain. The air intake line for all the fixtures is called a vent stack, not a stack vent (A and B are incorrect). The pipe into which all the soil and waste lines empty is the soil stack (C is incorrect).

43. **C**   In acoustics, both the reverberation time and the loudness in an enclosed space are affected by the absorptivity of the surfaces in the space. Acoustical absorptivity is not the same as thermal absorptivity, nor are they measured in the same units.

44. **B**   Insolation refers to heat gain from the sun. Heat energy transmitted by radiation could be from the sun, or from some other source, such as a fireplace (C is incorrect). Both A and D refer to *insulation*, not insolation.

45. **A**   The frequency of a sound is the number of cycles per second of the sound wave, and is perceived by the ear as pitch.

46. **D**   CLTD is the model which uses the equation $q_c = U(A) \Delta T$, but substitutes CLTD for ΔT. CLTD adjusts for radiation and time lag in the basic conduction formula.

47. **B**   Both sound intensity level (IL) and sound pressure level (SPL) are measured in decibels, where a decibel is the relative value of the sound intensity or pressure compared to a reference value, based on a logarithmic scale. The IL and SPL have nearly identical values.

48. C   The heat pump system and the unitary system can have a meter on each unit, thus permitting separate billing. The multizone system uses a separate mixing box at the plant for each zone. This does not lend itself to separate metering, since mixing boxes are nearly impossible to meter and there is only one plant serving all the zones.

49. A   A step down transformer is used to transform high voltage, such as that at a power line, to a lower voltage appropriate for building use.

50. D   Relative humidity is the ratio of the actual amount of water vapor in the air to the maximum amount that the air could contain at the same temperature. It is usually expressed as a percentage.

51. D   Choice I is not correct. The zone of the earth that supports human life (five miles into the earth's crust and five miles into the atmosphere) is an extremely fragile ecosystem. The biosphere that has evolved over millions of years, is dramatically affected by the growth of human activity in the last 150 years.

Choice II is not correct either. While innovative technologies are improving energy efficiency of some building systems, the vast majority of the built environment is energy inefficient.

Choice III is also not correct. Toxic substances have the tendency to expand and affect large areas. For example, the air above the Great Lakes contains evidence of DDT, a toxic pesticide banned in the USA decades ago. It was discovered that DDT is captured in the jet stream bringing toxic materials from far away continents that still use toxic pesticides

Choice IV is not correct. While recycling is helpful, it is just the beginning of the sustainable design process. The principles of sustainable design say that we need to have more building products that can be recycled and are biodegradable to create a more sustainable ecosystem.

52. D   All of these consultants (wetlands engineer, energy commissioner, landscape architect, and energy modeling engineer) might be necessary for the holistic approach to sustainable design. The landscape architect should have experience with local, native plant design.

53. A   I is correct. Solar shading, whether from landscaping or architectural elements, can regulate the insulation to increase winter light and reduce warm summer sunlight.

II is also correct. Urban heat island effect is the tendency of a building roof to absorb solar radiation during the day and then emit heat radiation during the evening. Roof systems with grass or light colored roofing material reduce the urban heat island effect.

III is not correct. Sustainable design encourages approaches that reduce the area allocated to parking.

IV is correct. The type, location, and size of building fenestration are a keys aspect of architectural design for sustainable projects.

INDEX

198   Mechanical and Electrical Systems

Comfort range or comfort zone, 26
Commission Internationale d'Eclairage
    (CIE), 85
Compartmentation, 149
Composting toilet, 138
Condenser, 54
Conduction, 6, 9
Conductivity (k), 6
Conduit, 99
Cones, 74
Constant volume, 58
Construction type, 148
Cooling load temperature differential, 14
Cooling tower, 55
Copper tubing, 130
Crack method, 11
Current, 89

D

Damper, 58
Darrieus, 46
Daylight, 82
Daylight factor method, 85
DBA, 109
DC motor, 97
Dead, 114
Decibel, 107
Deck temperature, 58
Declination, 28
Degree days, 9
Deluge system, 156
Demand surcharge, 102
Desiccant cooling, 44
Design day, 10
Dew point temperature, 34
Diffuser, 58
Diffusing, 75
Direct current, 90
Direct gain space, 38
Direct normal intensity (I_{dn}), 31
Distribution, 53
Domestic hot water, 42
Double duct, 58

Double envelope, 40
Downdraft, 54
Downfeed, 132
Drain back, 44
Drain down, 44
Dry bulb, 24
Dry standpipe, 153
Dual duct, 58
Duct sizing, 64

E

Earth sheltered, 41
Economizer cycle, 55
Efficacy, 78
Electrical metallic tubing (EMT), 99
Electromagnetic radiation, 73
Electrostatic filters, 62
Emergency lighting, 86
Emergency power source, 102
Emissivity, 3
Energy codes, 64
Enthalpy, 24
Equinoxes, 28
Equivalent circular diameter, 64
Equivalent spherical illumination (ESI), 82
Evaporator, 54
Exit lighting, 86
Exits, 150
Extinguishing media, 151

F

Fan coil, 60
Fan sizing, 64
Fibrous filter, 62
Filament, 78
Film coefficient (f_i), 6
Fins, 18, 30
Fire safety, 147
Fixture unit (FU), 136
Flashover, 152
Flat plate collector, 42
Flexible metal conduit, 99
Flow rate, 133